Unity 开发基础教程

主　　编　张一民　冯　丹

副主编　冯永明　齐安智　杨　洋　李晶晶

参　　编　李　洋　张福东　宋思佳

翻　　译　钱　瑞

北京理工大学出版社

BEIJING INSTITUTE OF TECHNOLOGY PRESS

内 容 提 要

Unity是实时3D互动内容创作和运营平台，包括游戏开发、美术、建筑、汽车设计、影视在内的所有创作者，借助Unity将创意变成现实。Unity平台提供一整套完善的软件解决方案，可用于创作、运营和变现任何实时互动的2D和3D内容，支持平台包括手机、平板电脑、PC、游戏主机、增强现实和虚拟现实设备。本书详细介绍了如何使用Unity进行开发，包括Unity历史、安装步骤等相关基础知识，以及如何使用C#语言构建脚本、输入系统、UI系统、射线、寻路系统、动画等相关内容。

本书可作为计算机相关专业设计开发类课程教材，也适合具备一些C#语言基础并且想快速入门Unity 3D游戏开发的人员阅读，同时本书内容对于零基础的新手开发者十分友好，可作为其学习参考资料。

图书在版编目（CIP）数据

Unity开发基础教程：汉英对照 / 张一民，冯丹主编.——北京：北京理工大学出版社，2023.1
ISBN 978-7-5763-1719-0

Ⅰ.①U… Ⅱ.①张… ②冯… Ⅲ.①游戏程序—程序设计—教材—汉、英 Ⅳ.①TP311.5

中国版本图书馆CIP数据核字（2022）第172359号

出版发行 / 北京理工大学出版社有限责任公司
社　　　址 / 北京市海淀区中关村南大街5号
邮　　　编 / 100081
电　　　话 /（010）68914775（总编室）
　　　　　　（010）82562903（教材售后服务热线）
　　　　　　（010）68944723（其他图书服务热线）
网　　　址 / http://www.bitpress.com.cn
经　　　销 / 全国各地新华书店
印　　　刷 / 河北鑫彩博图印刷有限公司
开　　　本 / 787毫米×1092毫米　1/16
印　　　张 / 17　　　　　　　　　　　　　　　责任编辑 / 钟　博
字　　　数 / 402千字　　　　　　　　　　　　文案编辑 / 钟　博
版　　　次 / 2023年1月第1版　2023年1月第1次印刷　责任校对 / 周瑞红
定　　　价 / 89.00元　　　　　　　　　　　　责任印制 / 王美丽

图书出现印装质量问题，请拨打售后服务热线，本社负责调换

前 言
PREFACE

    本教材根据"十三五"职业教育国家规划教材建设工作通知对编写教材新的要求，采取工作手册形式编写，内容反映新知识、新技术、新工艺、新方法，是相关专业高质量发展的体现。

    本书是教学中重要的知识补充，为提供简明易懂的技术知识，编者团队专门参加工作手册式教材设计与使用等培训，走访企业，与许多有经验的企业工作人员共同研究探讨，结合一线教学经验总结编写。

    本书共设计16个学习项目，双语（中英文）编写，每个学习项目含有学习性工作任务单、资讯单、计划单、决策单、实施单、检查单、评价单、教学引导文设计单、教学反馈单、分组单、教师实施计划单、成绩报告单12个表单，每学一个项目，可以对应填写项目中的各个表单，以使教材更加灵活，便于师生使用。在内容上，融入了职业元素，根据企业岗位的要求进行编排，通过项目带动学习任务，使学生能在理论上以够用为度，在实践中以会用为本，使教材与行业同步发展。

    本书配有精美的学习视频，读者可以自行扫码观看。

    由于编者水平有限，书中难免存在错误与疏漏，恳请广大读者批评指正。

编 者

Preface

This teaching material is based on the "13th Five Year Plan" national planning teaching material construction notice for vocational education and the new requirements for the compilation of teaching materials.We develop the compilation of work manual style teaching materials. The content of this teaching material reflects the new knowledge, new technology, new process, new methods, and the ability to improve the high–quality development of related majors.

The teaching materials are the most important supplement to knowledge in teaching, and the development of loose–leaf teaching materials can provide concise and easy–to–understand technical knowledge. For this reason, we specifically attended training on the design and use of loose–leaf and workbook–style teaching materials, visited enterprises and worked with many experienced enterprise staff to study the form of loose–leaf teaching materials, and finally co–edited "Unity Development Foundation" with front–line teachers who have rich teaching experience.

The book is designed with a total of sixteen learning projects, each of which contains learning task sheets, information sheets, planning sheets, decision sheets, implementation sheets, check sheets, evaluation sheets, teaching guide text design sheets, teaching feedback sheets, grouping sheets, teachers' implementation plan sheets, and achievement report sheets in twelve forms. The content of the course is integrated with vocational elements. In terms of content, it incorporates vocational elements and is written according to the requirements of corporate positions. Through project–driven learning tasks, students are able to use the material in theory to the extent that it is sufficient, and in practice to the extent that it is necessary, so that the material develops in tandem with the industry and makes the teaching materials and the industry develop synchronously.

The book comes with a exquisite learning video that can be downloaded by scanning a QR code.

Due to the author's limited level, there are inevitably mistakes in the book, imploring readers to criticize and correct.

Editors

目 录
CONTENTS

项目 4　如何在 Unity 中编写脚本

项目 5　Unity 中 GameObject 类

项目 6　Unity 中 Transform 类

项目 7　Prefab 预制体

项目 8　Collider 碰撞器

项目 9　Rigidbody 刚体

项目 10　Input 输入系统

项目 11　UI 系统

项目 12　射线 Ray

项目 13　寻路系统

项目 14　动画

项目 15　IK 反向运动学

项目 16　XML 文件

Contents

目 录

Project 1　Unity Basic Interface

Project 2　Unity Basic interface operation

Project 3　Inspector panel function, Camera component, Light component function introduction

Project 4　How to write scripts in Unity

Project 5　The Game Object class in Unity

Project 6　Transform class in Unity

Project 7　Prefab prefabricated body

Project 8　Collider

Project 9　Rigidbody

Project 10   Input System

Project 11   UI system

Project 12   Ray

Project 13   Pathfinding System

Project 14   Animations

项目1　Unity基础界面

1.1　项目表单

表 1-1　学习性工作任务单

学习场 K	Unity 基础界面					
学习情境 L	熟悉基础界面					
学习任务 M	进行简单操作			学时		4 学时(180 min)
典型工作过程描述	创建项目—了解基础界面—简单操作					
学习目标	1. 了解 Unity 2. 熟悉 Unity 基础界面 3. 可进行 Unity 简单操作					
任务描述	全面了解 Unity 基础界面并进行 Unity 简单操作。					
学时安排	资讯 20 min	计划 10 min	决策 10 min	实施 100 min	检查 20 min	评价 20 min
对学生的要求	1. 软件安装完成 2. 课前做好预习 3. 了解基础界面 4. 可进行 Unity 简单操作					
参考资料	1. 素材包 2. 微视频 3. PPT					

表 1-2 资讯单

学习场 K	Unity 基础界面		
学习情境 L	熟悉基础界面		
学习任务 M	进行简单操作	学时	20 min
典型工作过程描述	创建一个新的项目—全面了解基础界面—进行 Unity 的简单操作		
搜集资讯的方式	1. 教师讲解 2. 互联网查询 3. 同学交流		
资讯描述	查看教师提供的资料，获取信息，便于操作		
对学生的要求	1. 准备好学习用品及任务书 2. 课前做好预习 3. 熟悉 Unity 基础界面 4. 简单操作		
参考资料	1. 素材包 2. 微视频 3. PPT		

表 1-3 计划单

学习场 K	Unity 基础界面		
学习情境 L	熟悉基础界面		
学习任务 M	进行简单操作	学时	10 min
典型工作过程描述	创建一个新的项目—全面了解基础界面—进行 Unity 的简单操作		
计划制订的方式	同学间分组讨论		
序号	工作步骤	注意事项	
1	创建一个新的项目		
2	全面了解基础界面		
3	进行 Unity 的简单操作		
计划评价	班级		第____组	组长签字	
	教师签字		日期		
	评语：				

表 1-4 决策单

学习场 K	Unity 基础界面			
学习情境 L	熟悉基础界面			
学习任务 M	进行简单操作		学时	10 min
典型工作过程描述	创建一个新的项目—全面了解基础界面—进行 Unity 的简单操作			
计划对比				
序号	计划的可行性	计划的经济性	计划的可操作性	计划的实施难度	综合评价
1					
2					
3					
4					
5					
6					
7					
8					
9					
10					

	班级		第_____组	组长签字	
	教师签字		日期		
决策评价	评语：				

表 1-5　实施单

学习场 K	Unity 基础界面			
学习情境 L	熟悉基础界面			
学习任务 M	进行简单操作		学时	100 min
典型工作过程描述	创建一个新的项目—全面了解基础界面—进行 Unity 的简单操作			

序号	实施步骤	注意事项
1	创建一个新的项目	新建项目
2	全面了解基础界面	视图和菜单栏
3	进行 Unity 的简单操作	物体的简单操作，例如移动或缩放等

实施说明：

1. 启动 UnityHub 程序后，需要创建一个项目
2. 全面了解基础界面
3. 根据了解的内容，进行 Unity 简单操作

实施评价	班级		第_____组	组长签字	
	教师签字		日期		
	评语：				

表 1-6　检查单

学习场 K	Unity 基础界面			
学习情境 L	熟悉基础界面			
学习任务 M	进行简单操作		学时	20 min
典型工作过程描述	创建一个新的项目—全面了解基础界面—进行 Unity 的简单操作			

序号	检查项目	检查标准	学生自查	教师检查
1	资讯环节	获取相关信息情况		
2	计划环节	设计物体简单操作		
3	实施环节	创建物体进行设置		
4	检查环节	各个环节逐一检查		

检查评价	班级		第_____组	组长签字	
	教师签字		日期		
	评语：				

表 1-7　评价单

学习场 K	Unity 基础界面			
学习情境 L	熟悉基础界面			
学习任务 M	进行简单操作		学时	20 min
典型工作过程描述	创建一个新的项目—全面了解基础界面—进行 Unity 的简单操作			
评价项目	评价子项目	学生自评	组内评价	教师评价
资讯环节	1. 听取教师讲解 2. 互联网查询情况 3. 同学交流情况			
计划环节	1. 查询资料情况 2. 设计物体简单操作			
实施环节	1. 学习态度 2. 使用软件的熟练程度			
最终结果	综合情况			

评价	班级		第_____组	组长签字	
	教师签字		日期		
	评语：				

表 1-8　教学引导文设计单

学习场 K	Unity 基础界面	学习情境 L	熟悉基础界面布局	参照系	信息工程学院
		学习任务 M	进行简单操作		
普适性 工作过程 典型工作 过程	资讯	计划	决策	实施	检查	评价
分析基础 界面布局	教师讲解	同学分组讨论	计划的可行性	了解基础 界面	获取相关 信息情况	评价学习 态度
熟悉基础 界面布局	自行掌握	熟悉布局	计划的经济性	设置布局方式	检查参数	评价学生 的熟悉度
进行简单操作	根据掌握知识 进行物体的 简单操作	设计简单操作	计划的 实施难度	创建物体并进 行简单的操作	检查参数及组 件是否运用正确	软件熟练 程度
保存项目文件	了解项目 文件的格式	了解项目 文件的格式	综合评价	保存项目文件	检查项目 文件的格式	评价项目

表 1-9 教学反馈单(学生反馈)

学习场 K	Unity 基础界面		
学习情境 L	熟悉基础界面		
学习任务 M	进行简单操作	学时	4 学时(180 min)
典型工作过程描述	创建一个新的项目—全面了解基础界面—进行 Unity 的简单操作		

调查项目	序号	调查内容	理由描述
	1	资讯环节	
	2	计划环节	
	3	实施环节	
	4	检查环节	

您对本次课程教学的改进意见：

调查信息	被调查人姓名		调查日期	

表 1-10 分组单

学习场 K	Unity 基础界面		
学习情境 L	熟悉基础界面		
学习任务 M	进行简单操作	学时	4 学时(180 min)
典型工作过程描述	创建一个新的项目—全面了解基础界面—进行 Unity 的简单操作		

分组情况	组别	组长		组员
	1			
	2			
	3			
	4			
	5			
	6			
	7			
	8			
分组说明				

班级		教师签字		日期	

表 1-11 教师实施计划单

学习场 K	Unity 基础界面			
学习情境 L	熟悉基础界面			
学习任务 M	进行简单操作		学时	4 学时(180 min)
典型工作过程描述	创建一个新的项目—全面了解基础界面—进行 Unity 的简单操作			

序号	工作与学习步骤	学时	使用工具	地点	方式	备注
1	资讯情况	20 min	互联网			
2	计划情况	10 min	计算机			
3	决策情况	10 min	计算机			
4	实施情况	100 min	Unity			
5	检查情况	20 min	计算机			
6	评价情况	20 min	课程伴侣			

班级		教师签字		日期	

表 1-12 成绩报告单

班级Unity 基础界面学习场(课程)成绩报告单														
学习场 K	Unity 基础界面													
学习情境 M	熟悉基础界面													
典型工作过程描述	进行简单操作							学时			4 学时(180 min)			

序号	姓名	第一个学习任务				第二个学习任务				第 N 个学习任务				总评
		自评 ×%	互评 ×%	教师评 ×%	合计	自评 ×%	互评 ×%	教师评 ×%	合计	自评 ×%	互评 ×%	教师评 ×%	合计	
1														
2														
3														
4														
5														
6														
7														
8														
9														
10														
11														
12														
13														
14														
15														
16														

序号	姓名	第一个学习任务				第二个学习任务				第N个学习任务				总评
		自评 ×%	互评 ×%	教师评 ×%	合计	自评 ×%	互评 ×%	教师评 ×%	合计	自评 ×%	互评 ×%	教师评 ×%	合计	
17														
18														
19														
20														
21														
22														
23														
24														
25														
26														
27														
28														
班级		教师签字								日期				

1.2　理论指导

1.2.1　Unity 基础界面布局

打开 Unity 编辑器，可以看到很多视图，包括 Hierarchy(层次视图)、Scene(场景视图)、Project(项目视图)、Inspector(检视视图)、Game(游戏视图)，这些视图之间有着非常紧密的联系，可以更加清晰地显示出整个游戏工程的层次、架构与概念(图 1-1)。

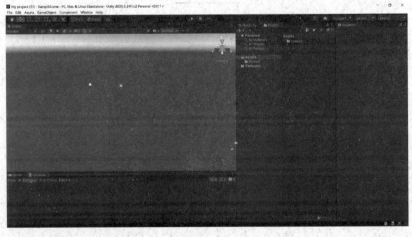

**Unity 基础界面
布局及按钮功能**

图 1-1

1. Hierarchy 视图（层级视图）

Hierarchy 视图主要用于存放游戏场景中具体的游戏对象和对象之间的层级关系，Hierarchy 视图的左上角有个"Create"，通过单击其中的选项可以创建新的游戏对象。在 Hierarchy 视图中可建立游戏对象之间的父子关系（图 1-2）。

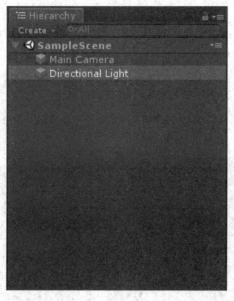

图 1-2

2. Scene 视图（场景视图）

Scene 视图主要用于编辑整个游戏世界，将各种美术资源以及系统自带的资源（Gizmes），在场景视图中进行搭建，其中还包括灯光、模型和特效等。在 Hierarchy 视图中选择立方体对象，再在 Scene 视图中按快捷键 F 来近距离查看该游戏对象（图 1-3）。

图 1-3

3. Project 视图（项目视图）

Project 视图用于存放游戏设计中用到的所有资源文件和脚本。Project 视图的左上角有个"Create"，通过单击其中的选项可以创建新的游戏资源（图1-4）。

4. Inspector 视图（检视视图）

Inspector 视图用于呈现游戏对象、游戏资源、游戏设置以及展示描述信息和属性的地方。在此视图中，会有选择组件的描述和该组件描述的所有参数，并且有一部分的组件参数可动态修改（图1-5）。

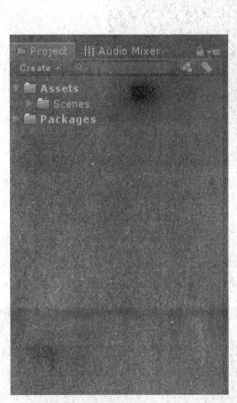

图 1-4   图 1-5

5. Game 视图（游戏视图）

Game 视图是最终游戏发布后展示在屏幕中的效果，就是游戏运行起来的结果展示，屏幕展示的内容完全取决于 Hierarchy 视图中摄像机照射的部分（图1-6）。

1.2.2 顶部菜单栏

与其他软件相同，顶部菜单栏包含了 Unity 主要的功能和设置，共有 7 个菜单栏。

图 1-6

1. File 菜单

File(文件)菜单主要用于创建、储存场景、项目(图 1-7)。

2. Edit 菜单

Edit(编辑)菜单主要用于场景中各个对象的编辑设置(图 1-8)。

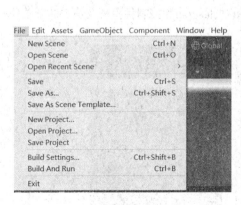

图 1-7

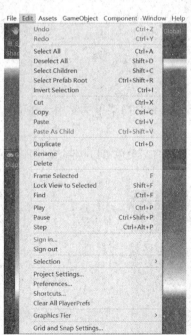

图 1-8

3. Assets 菜单

Assets(资源)菜单可以导入或导出所用的资源包，是 Unity 提供的用来管理游戏资源所需要的资源包(图1-9)。

4. Game Object 菜单

Game Object(游戏)菜单主要用于在场景中添加游戏对象，以及对对象的设置(图1-10)。

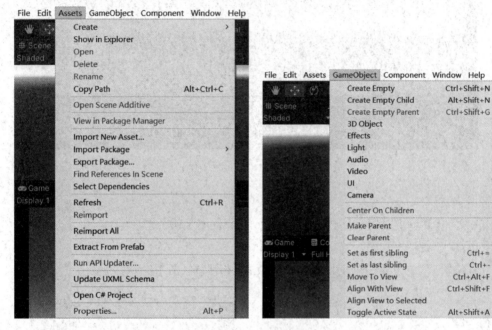

图1-9   图1-10

5. Component 菜单

Component(组件)菜单主要用于在项目制作过程中为游戏物体添加组件或属性(图1-11)。

6. Window 菜单

Window(窗口)菜单可以控制整个编辑器的页面布局和控制各个视图窗口的开关(图1-12)。

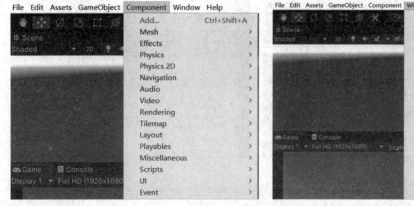

图1-11   图1-12

7. Help 菜单

Help(帮助)菜单用于帮助用户学习和掌握 Unity(图 1-13)。

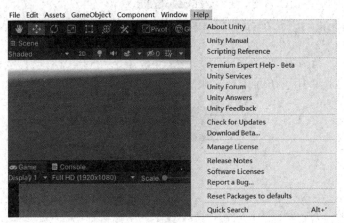

图 1-13

1.2.3 工具栏

工具栏包含 7 个基本控件。每个控件与 Editor 的不同部分相关。

1. 变换组件工具

变换组件工具用于操作 Scene 视图,主要用于实现对游戏对象的控制,包括位置、旋转、缩放等(图 1-14)。

图 1-14

2. 变换辅助图标开关

变换辅助图标开关影响 Scene 视图显示效果,主要是对游戏对象进行位置变换操作(图 1-15)。

图 1-15

3. 播放/暂停/步进按钮

播放/暂停/步进按钮用于处理 Game 视图,主要用于对项目的控制,项目的播放、暂停、步进(图 1-16)。

图 1-16

4. 云按钮

云按钮用于打开 Unity Services 窗口(图 1-17)。

图 1-17

5. Account 下拉选单

Account 下拉选单用于访问 Unity 账户（图 1-18）。

图 1-18

6. Layers 分层下拉选单

Layers 分层下拉选单控制 Scene 视图中显示的对象（图 1-19）。

图 1-19

7. Layout 布局下拉选单

Layout 布局下拉选单控制所有视图的布局（图 1-20）。

图 1-20

项目2　Unity基础界面操作

2.1　项目表单

表 2-1　学习性工作任务单

学习场 K	Unity 基础界面操作					
学习情境 L	熟悉基础界面操作					
学习任务 M	进行复杂操作			学时	4 学时(180 min)	
典型工作过程描述	创建项目—了解基础界面操作—进行复杂操作					
学习目标	1. 了解 Unity 2. 熟悉 Unity 基础界面操作 3. 进行复杂操作					
任务描述	熟悉 Unity 基础界面操作并进行 Unity 复杂操作					
学时安排	资讯 20 min	计划 10 min	决策 10 min	实施 100 min	检查 20 min	评价 20 min
对学生的要求	1. 安装好软件 2. 课前做好预习 3. 熟悉基础界面操作 4. Unity 复杂操作					
参考资料	1. 素材包 2. 微视频 3. PPT					

表 2-2　资讯单

学习场 K	Unity 基础界面操作		
学习情境 L	熟悉基础界面操作		
学习任务 M	进行复杂操作	学时	20 min
典型工作过程描述	创建项目—了解基础界面操作—进行复杂操作		
搜集资讯的方式	1. 教师讲解 2. 互联网查询 3. 同学间交流		
资讯描述	查看教师提供的资料，获取信息，便于操作		
对学生的要求	1. 软件安装完成 2. 课前做好预习 3. 熟悉基础界面操作 4. 进行 Unity 复杂操作		
参考资料	1. 素材包 2. 微视频 3. PPT		

表 2-3　计划单

学习场 K	Unity 基础界面操作		
学习情境 L	熟悉基础界面操作		
学习任务 M	进行复杂操作	学时	10 min
典型工作过程描述	创建项目—了解基础界面操作—进行复杂操作		
计划制订的方式	同学间分组讨论		

序号	工作步骤	注意事项
1	创建一个新的项目	
2	熟悉基础界面操作	
3	进行 Unity 的复杂操作	

计划评价	班级		第___组	组长签字	
	教师签字		日期		
	评语：				

表 2-4 决策单

学习场 K	Unity 基础界面操作			
学习情境 L	熟悉基础界面操作			
学习任务 M	进行复杂操作		学时	10 min
典型工作过程描述	创建项目—了解基础界面操作—进行复杂操作			
计划对比				
序号	计划的可行性	计划的经济性	计划的可操作性	计划的实施难度	综合评价
1					
2					
3					
4					
5					
6					
7					
8					
9					
10					

	班级		第____组	组长签字	
	教师签字		日期		
决策评价	评语:				

表 2-5 实施单

学习场 K	Unity 基础界面操作		
学习情境 L	熟悉基础界面操作		
学习任务 M	进行复杂操作	学时	100 min
典型工作过程描述	创建项目—了解基础界面操作—进行复杂操作		
序号	实施步骤	注意事项	
1	创建一个新的项目	新建项目	
2	全面了解基础界面操作	界面操作	
3	进行 Unity 的复杂操作	物体的复杂操作,例如移动或缩放或建立父子关系等	

实施说明:

1. 启动 UnityHub 程序后,需要创建一个项目

2. 全面了解基础界面操作

3. 根据所掌握的知识进行 Unity 复杂操作

	班级		第____组	组长签字	
	教师签字		日期		
实施评价	评语:				

表 2-6　检查单

学习场 K	Unity 基础界面操作			
学习情境 L	熟悉基础界面操作			
学习任务 M	进行复杂操作		学时	20 min
典型工作过程描述	创建项目—了解基础界面操作—进行复杂操作			
序号	检查项目	检查标准	学生自查	教师检查
1	资讯环节	获取相关信息情况		
2	计划环节	设计物体复杂操作		
3	实施环节	创建物体进行设置		
4	检查环节	各个环节逐一检查		
检查评价	班级		第＿＿组	组长签字	
	教师签字		日期		
	评语：				

表 2-7　评价单

学习场 K	Unity 基础界面操作			
学习情境 L	熟悉基础界面操作			
学习任务 M	进行复杂操作		学时	20 min
典型工作过程描述	创建项目—了解基础界面操作—进行复杂操作			
评价项目	评价子项目	学生自评	组内评价	教师评价
资讯环节	1. 听取教师讲解 2. 互联网查询情况 3. 同学交流情况			
计划环节	1. 查询资料情况 2. 设计物体复杂操作			
实施环节	1. 学习态度 2. 使用软件的熟练程度			
最终结果	综合情况			
评价	班级		第＿＿组	组长签字	
	教师签字		日期		
	评语：				

表 2-8 教学引导文设计单

学习场 K	Unity 基础界面操作	学习情境 L	熟悉基础界面操作	参照系		信息工程学院
		学习任务 M	进行复杂操作			
普适性工作过程　　典型工作过程	资讯	计划	决策	实施	检查	评价
分析基础界面布局操作	教师讲解	同学分组讨论	计划的可行性	熟悉基础界面布局操作	获取相关信息情况	评价学习态度
熟悉基础界面布局操作	自行掌握	熟悉布局操作	计划的经济性	设置操作方式	检查参数	评价学生的熟悉度
进行简单操作	根据掌握知识进行物体的复杂操作	设计复杂操作	计划的实施难度	创建物体并进行简单的操作	检查参数及组件是否运用正确	软件熟练程度
保存项目文件	了解项目文件的格式	了解项目文件的格式	综合评价	保存项目文件	检查项目文件的格式	评价项目

表 2-9 教学反馈单(学生反馈)

学习场 K	Unity 基础界面操作			
学习情境 L	熟悉基础界面操作			
学习任务 M	进行复杂操作		学时	4 学时(180 min)
典型工作过程描述	创建项目—了解基础界面操作—进行复杂操作			
调查项目	序号	调查内容		理由描述
	1	资讯环节		
	2	计划环节		
	3	实施环节		
	4	检查环节		

您对本次课程教学的改进意见:

调查信息	被调查人姓名		调查日期

表 2-10 分组单

学习场 K	Unity 基础界面操作				
学习情境 L	熟悉基础界面操作				
学习任务 M	进行复杂操作			学时	4 学时(180 min)
典型工作过程描述	创建项目—了解基础界面操作—进行复杂操作				
分组情况	组别	组长		组员	
	1				
	2				
	3				
	4				
	5				
	6				
	7				
	8				
分组说明					
班级		教师签字		日期	

表 2-11 教师实施计划单

学习场 K	Unity 基础界面操作				
学习情境 L	熟悉基础界面操作				
学习任务 M	进行复杂操作		学时	4 学时(180 min)	
典型工作过程描述	创建项目—了解基础界面操作—进行复杂操作				
序号	工作与学习步骤	学时	使用工具	地点	方式	备注
1	资讯情况	20 min	互联网			
2	计划情况	10 min	计算机			
3	决策情况	10 min	计算机			
4	实施情况	100 min	Unity			
5	检查情况	20 min	计算机			
6	评价情况	20 min	课程伴侣			
班级		教师签字		日期		

表 2-12 成绩报告单

_____班级 Unity 基础界面操作学习场(课程)成绩报告单														
学习场 K	Unity 基础界面操作													
学习情境 M	熟悉基础界面操作													
典型工作过程描述	进行复杂操作							学时	4 学时(180 min)					
序号	姓名	第一个学习任务				第二个学习任务				第 N 个学习任务				总评
		自评 ×%	互评 ×%	教师评 ×%	合计	自评 ×%	互评 ×%	教师评 ×%	合计	自评 ×%	互评 ×%	教师评 ×%	合计	
1														

续表

序号	姓名	第一个学习任务				第二个学习任务				第N个学习任务				总评
		自评×%	互评×%	教师评×%	合计	自评×%	互评×%	教师评×%	合计	自评×%	互评X%	教师评×%	合计	
2														
3														
4														
5														
6														
7														
8														
9														
10														
11														
12														
13														
14														
15														
16														
17														
18														
19														
20														
21														
22														
23														
24														
25														
26														
27														
28														
班级		教师签字						日期						

2.2　理论指导

Unity 基础界面操作

Unity 基础界面操作，有以下几点。

1. Scene 视图操作

Scene 视图具有一组可用于快速有效移动的导航控件。

2. 场景视图辅助图标

场景视图辅助图标（Scene Gizmo）位于 Scene 视图的右上角。此控件用于显示 Scene 视图摄像机的当前方向，并允许快速修改视角和投影模式（图 2-1）。

3. 飞行模式

图 2-1

（1）单击并按住鼠标右键。

（2）使用鼠标移动视图，并按 W/A/S/D 键向左/向右/向前/向后移动，按 Q 和 E 键向上和向下移动。

（3）按住 Shift 键可以加快移动速度。

4. 游戏对象移动、旋转、缩放、矩形变换、变换

工具栏中的第一个工具手形工具（Hand Tool）用于平移场景，快捷键为 Q。选择手形工具并按住 Alt 键可以旋转当前的场景视角。另外，按住 Alt 键和鼠标右键左右拖动可以缩放和拉近场景，鼠标的滚轮也可以实现相同的效果。移动（Move）、旋转（Rotate）、缩放（Scale）、矩形变换（Rect Transform）和变换（Transform）工具用于编辑各个游戏对象。要更改游戏对象的变换组件，请使用鼠标操纵任何辅助图标轴，或直接在 Inspector 视图中的变换组件的数字字段中输入值。也可以使用快捷键选择变换模式：W 表示移动，E 表示旋转，R 表示缩放，T 表示矩形变换，Y 表示变换（图 2-2）。

（1）移动。在移动辅助图标的中心，有 3 个小方块可用于在单个平面内拖动游戏对象（意味着可一次移动两个轴，而第三个保持静止），同时还有 3 个轴向，代表着可以向任意轴移动（图 2-3）。

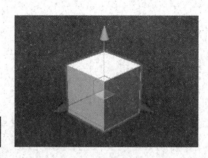

图 2-2　　　　　　　　　　　　　　　图 2-3

（2）旋转。选择旋转工具后，通过单击并拖动围绕游戏对象显示的线框球体辅助图标的轴来更改游戏对象的旋转（图 2-4）。

（3）缩放。使用缩放工具，可通过单击并拖动辅助图标中心的立方体，在所有轴上均匀地重新缩放游戏对象(图2-5)。

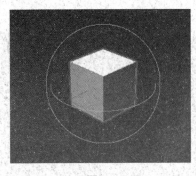
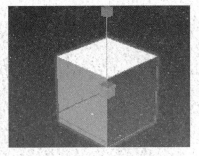

图2-4　　　　　　　　　　　　　　图2-5

(4)矩形变换。矩形变换通常用于定位2D元素(如精灵或UI元素),但也可用于操作3D游戏对象。此工具将移动、缩放和旋转功能整合到了同一个辅助图标中,具体操作如下:

①在矩形辅助图标中单击并拖动可移动游戏对象。

②单击并拖动矩形辅助图标的任何角或边可缩放游戏对象。

③拖动某条边可沿一个轴缩放游戏对象。

④拖动某个角可在两个轴上缩放游戏对象。

⑤要旋转游戏对象,请将光标放在矩形的某个角之外,光标变为显示旋转图标,单击并从此区域拖动可旋转游戏对象(图2-6)。

(5)变换。变换工具组合了移动、旋转和缩放工具。该工具的辅助图标提供了用于移动和旋转的控制柄(图2-7)。

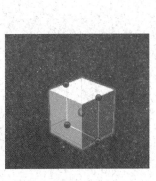

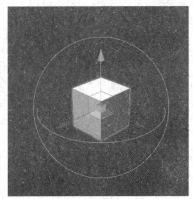

图2-6　　　　　　　　　　　　　　图2-7

5. 父子物体

(1)一个游戏对象只能有一个父物体。

(2)一个游戏对象可以有无数个子物体。

(3)当对一个子物体进行位移、旋转、缩放的时候,不会对父物体产生影响。原因:子物体的位置、旋转和缩放都是相对于父物体的,无论父物体怎么移动,相对父物体的位置是不变的计算机。

（4）当对一个父物体进行位移、旋转、缩放的时候，会对所有的子物体产生影响（图 2-8）。

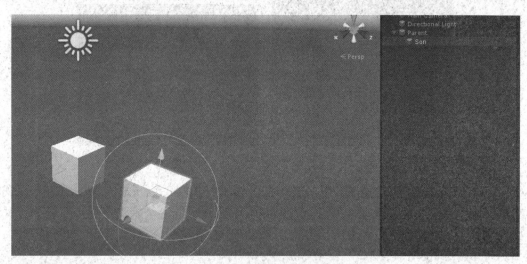

图 2-8

项目3　Inspector面板功能、Camera组件、Light组件功能介绍

3.1　项目表单

表 3-1　学习性工作任务单

学习场 K	Inspector 面板功能、Camera 组件、Light 组件功能介绍					
学习情境 L	熟悉面板功能与组件功能					
学习任务 M	组件与面板功能的运用			学时		4 学时(180 min)
典型工作过程描述	创建项目—熟悉面板功能与组件功能—组件与面板功能的运用					
学习目标	1. 了解面板功能与组件功能 2. 熟悉面板功能与组件功能 3. 组件与面板功能的运用					
任务描述	熟悉面板功能与组件功能并进行组件与面板功能的运用					
学时安排	资讯 20 min	计划 10 min	决策 10 min	实施 100 min	检查 20 min	评价 20 min
对学生的要求	1. 安装好软件 2. 课前做好预习 3. 熟悉面板功能与组件功能 4. 进行组件与面板功能的运用					
参考资料	1. 素材包 2. 微视频 3. PPT					

表3-2 资讯单

学习场 K	Inspector 面板功能、Camera 组件、Light 组件功能介绍		
学习情境 L	熟悉面板功能与组件功能		
学习任务 M	组件与面板功能的运用	学时	20 min
典型工作过程描述	创建项目—熟悉面板功能与组件功能—组件与面板功能的运用		
搜集资讯的方式	1. 教师讲解 2. 互联网查询 3. 同学交流		
资讯描述	查看教师提供的资料,获取信息,便于运用		
对学生的要求	1. 软件安装完成 2. 课前做好预习 3. 熟悉面板功能与组件功能 4. 组件与面板功能的运用		
参考资料	1. 素材包 2. 微视频 3. PPT		

表3-3 计划单

学习场 K	Inspector 面板功能、Camera 组件、Light 组件功能介绍			
学习情境 L	熟悉面板功能与组件功能			
学习任务 M	组件与面板功能的运用		学时	10 min
典型工作过程描述	创建项目—熟悉面板功能与组件功能—组件与面板功能的运用			
计划制订的方式	同学间分组讨论			
序号	工作步骤		注意事项	
1	创建一个新的项目			
2	熟悉面板功能与组件功能			
3	组件与面板功能的运用			
计划评价	班级		第____组	组长签字
	教师签字		日期	
	评语:			

表 3-4　决策单

学习场 K	Inspector 面板功能、Camera 组件、Light 组件功能介绍			
学习情境 L	熟悉面板功能与组件功能			
学习任务 M	组件与面板功能的运用		学时	10 min
典型工作过程描述	创建项目—熟悉面板功能与组件功能—组件与面板功能的运用			
计划对比				
序号	计划的可行性	计划的经济性	计划的可操作性	计划的实施难度	综合评价
1					
2					
3					
4					
5					
6					
7					
8					
9					
10					

	班级		第＿＿＿组	组长签字	
	教师签字		日期		
决策评价	评语：				

表 3-5　实施单

学习场 K	Inspector 面板功能、Camera 组件、Light 组件功能介绍		
学习情境 L	熟悉面板功能与组件功能		
学习任务 M	组件与面板功能的运用	学时	100 min
典型工作过程描述	创建项目—熟悉面板功能与组件功能—组件与面板功能的运用		
序号	实施步骤	注意事项	
1	创建一个新的项目	新建项目	
2	熟悉面板功能与组件功能	面板功能与组件功能	
3	组件与面板功能的运用	组件与面板功能的灵活运用	

实施说明：

1. 启动 UnityHub 程序后，需要创建一个项目
2. 熟悉面板功能与组件功能
3. 组件与面板功能的运用

	班级		第＿＿＿组	组长签字	
	教师签字		日期		
实施评价	评语：				

表 3-6 检查单

学习场 K	Inspector 面板功能、Camera 组件、Light 组件功能介绍			
学习情境 L	熟悉面板功能与组件功能			
学习任务 M	组件与面板功能的运用		学时	20 min
典型工作过程描述	创建项目—熟悉面板功能与组件功能—组件与面板功能的运用			
序号	检查项目	检查标准	学生自查	教师检查
1	资讯环节	获取相关信息情况		
2	计划环节	熟悉面板功能与组件功能		
3	实施环节	组件与面板功能的运用		
4	检查环节	各个环节逐一检查		
检查评价	班级		第＿＿组	组长签字
	教师签字		日期	
	评语：			

表 3-7 评价单

学习场 K	Inspector 面板功能、Camera 组件、Light 组件功能介绍			
学习情境 L	熟悉面板功能与组件功能			
学习任务 M	组件与面板功能的运用		学时	20 min
典型工作过程描述	创建项目—熟悉面板功能与组件功能—组件与面板功能的运用			
评价项目	评价子项目	学生自评	组内评价	教师评价
资讯环节	1. 听取教师讲解 2. 互联网查询情况 3. 同学交流情况			
计划环节	1. 查询资料情况 2. 熟悉面板功能与组件功能			
实施环节	1. 学习态度 2. 使用软件的熟练程度			
最终结果	综合情况			
评价	班级		第＿＿组	组长签字
	教师签字		日期	
	评语：			

表 3-8　教学引导文设计单

学习场 K	Inspector 面板功能、Camera 组件、Light 组件功能介绍	学习情境 L	熟悉面板功能与组件功能	参照系	信息工程学院
		学习任务 M	组件与面板功能的运用		

典型工作过程 ＼ 普适性工作过程	资讯	计划	决策	实施	检查	评价
分析面板功能与组件功能	教师讲解	同学分组讨论	计划的可行性	了解面板功能与组件功能	获取相关信息情况	评价学习态度
熟悉面板功能与组件功能	自行掌握	熟悉面板功能与组件功能	计划的经济性	设置操作方式	检查参数	评价学生的熟悉度
进行运用	组件与面板功能的运用	设计操作	计划的实施难度	组件与面板功能的运用	检查参数及组件是否运用正确	软件熟练程度
保存项目文件	了解项目文件的格式	了解项目文件的格式	综合评价	保存项目文件	检查项目文件的格式	评价项目

表 3-9　教学反馈单(学生反馈)

学习场 K	Inspector 面板功能、Camera 组件、Light 组件功能介绍			
学习情境 L	熟悉面板功能与组件功能			
学习任务 M	组件与面板功能的运用		学时	4 学时(180 min)
典型工作过程描述	创建项目—熟悉面板功能与组件功能—组件与面板功能的运用			
调查项目	序号	调查内容	理由描述	
	1	资讯环节		
	2	计划环节		
	3	实施环节		
	4	检查环节		
您对本次课程教学的改进意见：				
调查信息	被调查人姓名		调查日期	

表 3-10 分组单

学习场 K	Inspector 面板功能、Camera 组件、Light 组件功能介绍			
学习情境 L	熟悉面板功能与组件功能			
学习任务 M	组件与面板功能的运用		学时	4 学时(180 min)
典型工作过程描述	创建项目—熟悉面板功能与组件功能—组件与面板功能的运用			
分组情况	组别	组长	组员	
	1			
	2			
	3			
	4			
	5			
	6			
	7			
	8			
分组说明				
班级		教师签字	日期	

表 3-11 教师实施计划单

学习场 K	Inspector 面板功能、Camera 组件、Light 组件功能介绍					
学习情境 L	熟悉面板功能与组件功能					
学习任务 M	组件与面板功能的运用		学时	4 学时(180 min)		
典型工作过程描述	创建项目—熟悉面板功能与组件功能—组件与面板功能的运用					
序号	工作与学习步骤	学时	使用工具	地点	方式	备注
1	资讯情况	20 min	互联网			
2	计划情况	10 min	计算机			
3	决策情况	10 min	计算机			
4	实施情况	100 min	Unity			
5	检查情况	20 min	计算机			
6	评价情况	20 min	课程伴侣			
班级		教师签字			日期	

表 3-12 成绩报告单

班级Inspector 面板功能、Camera 组件、Light 组件功能介绍学习场(课程)成绩报告单														
学习场 K		Inspector 面板功能、Camera 组件、Light 组件功能介绍												
学习情境 M		熟悉面板功能与组件功能												
典型工作过程描述		组件与面板功能的运用							学时			4 学时(180 min)		
序号	姓名	第一个学习任务				第二个学习任务				第 N 个学习任务				总评
		自评 ×%	互评 ×%	教师评 ×%	合计	自评 ×%	互评 ×%	教师评 ×%	合计	自评 ×%	互评 ×%	教师评 ×%	合计	
1														
2														
3														
4														
5														
6														
7														
8														
9														
10														
11														
12														
13														
14														
15														
16														
17														
18														
19														
20														
21														
22														
23														
24														
25														
26														
27														
28														
班级			教师签字							日期				

3.2 理论指导

Inspector 面板功能、Camera 组件、Light 组件功能介绍如下：

3.2.1 Inspector 视图

Unity Editor 中的项目由多个游戏对象组成，而这些游戏对象包含脚本、声音、网格和其他图形元素（如光源）。Inspector 窗口（有时称为"Inspector"）显示有关当前所选游戏对象的详细信息，包括所有附加的组件及其属性，并允许修改场景中的游戏对象的功能（图 3-1）。

Inspector 面板功能

图 3-1

1. 检查游戏对象与脚本变量

使用 Inspector 可以查看和编辑 Unity Editor 中绝大多数的内容（包括物理游戏元素，如游戏对象、资源和材质）的属性和设置，以及 Editor 内的设置和偏好设置。在 Hierarchy 视图或 Scene 视图中选择游戏对象时，Inspector 将显示该游戏对象的所有组件和材质的属性，使用 Inspector 可以编辑这些组件和材质的设置（图 3-2）。

2. 检查资源

在 Project 窗口中选择资源后，Inspector 将显示关于如何导入资源和在运行时（游戏在 Editor 中或已发布的版本中运行）使用该资源的设置。每种类型的资源都有一组不同的

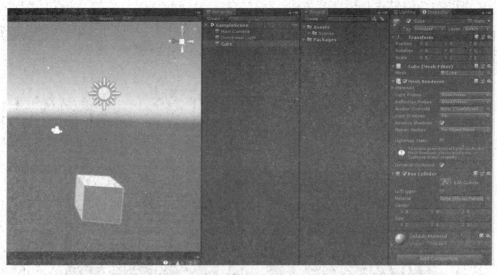

图 3-2

设置。例如材质资源如图 3-3 所示。

3. 游戏对象图标(Select Icon)

将图标分配给游戏对象后，图标将在 Scene 视图中显示在该游戏对象(以及之后的任何重复项)的上方。还可以将图标分配给预制件，从而将图标应用于场景中该预制件的所有实例(图 3-4)。

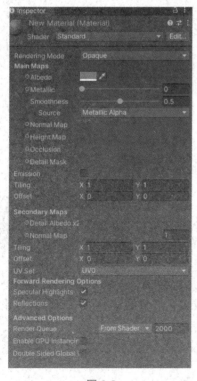

图 3-3

图 3-4

4. 显示与隐藏游戏物体

在 Unity 中创建一个 Game Object 放在 Hierarchy 视图上，如果要隐藏该物体，可以通过 Inspector 视图来设置，在最上面的选项，取消选中则隐藏物体(图 3-5)。

5. 设置静态

通过单击右上角"Static"按钮可以将物体设置为静态(图 3-6)

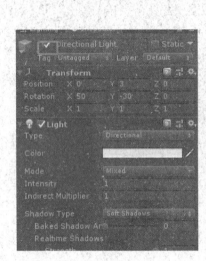

图 3-5　　　　　　　　　　　　　　　　　　图 3-6

6. 设置标签与层级

Tag 和 Layer 分别表示 Unity 引擎里面的标签和层，它们都是用来对 Game Object 进行标识的属性，Tag 常用于单个 Game Object，Layer 常用于一组的 Game Object。添加 Tag 和 Layer 的操作如下：

执行"Edit"→"Project Settings"→"Tags and Layers"命令来打开设置面板。也可以在下拉菜单中直接添加(图 3-7)。

7. 组件简介

一个游戏对象(Game Object)包含多个组件。一个组件代表一个功能(图 3-8)。

3.2.2　Camera 组件功能介绍

摄像机(Camera)是为玩家捕捉和展示世界的设备。通过自定义和操纵摄像机，可以让游戏呈现出真正的独特性。在场景中可拥有无限数量的摄像机。这些摄

图 3-7

像机可设置为以任何顺序在屏幕上任何位置或仅在屏幕的某些部分进行渲染(图 3-9)。

摄像机属性介绍：

(1)Clear Flags：确定将清除屏幕的哪些部分。此属性方便了使用多个摄像机来绘制不同游戏元素。

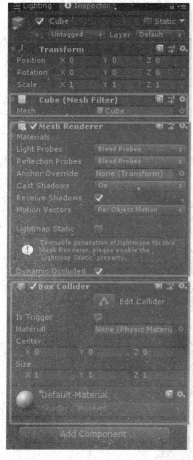

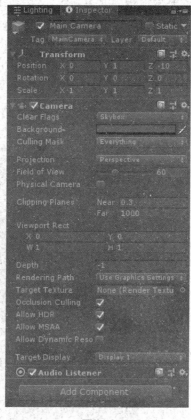

Camera 组件
功能介绍

图 3-8　　　　　　　　　　　　图 3-9

（2）Background：在绘制视图中的所有元素之后，但没有天空盒的情况下，应用于剩余屏幕部分的颜色。

（3）Culling Mask：选择所要显示的 Layer。

（4）Projection：

① Perspective：透视，摄像机将用透视的方式来渲染游戏对象。

② Field of View：视野范围，用于控制摄像机的视角宽度以及纵向的角度尺寸。

③ Orthographic：正交，摄像机将用无透视的方式来渲染游戏对象。

④ Size：大小，用于控制正交模式摄像机的视口大小。

（5）Clipping Planes：摄像机开始渲染与停止渲染之间的距离。

① Near：近点，摄像机开始渲染的最近的点。

② Far：远点，摄像机开始渲染的最远的点。

（6）Viewport Rect：用 4 个数值来控制摄像机的视图将绘制在屏幕的位置和大小，使用的是屏幕坐标系，数值为 0～1。坐标系原点在左下角。

X：距离 x 轴的位置；

Y：距离 y 轴的位置；

W：Weight 游戏对象所渲染的图像宽度；

H：Height 游戏对象所渲染的图像高度。

（7）Depth：用于控制摄像机的渲染顺序，较大值的摄像机将被渲染在较小值的摄像机之上。

（8）Target Texture：目标图像，可以自由设置在 Game 窗体下背景图片。

3.2.3　Light 组件功能介绍

Unity 中的光照主要由光源对象提供。光源决定对象的着色及其投射的阴影。因此，它们是图形渲染的基本部分(图 3-10)。

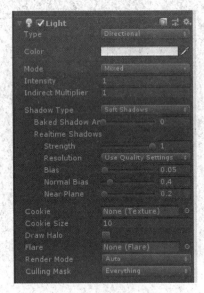

Light 组件功能介绍

图 3-10

1. Type 光源类型

Type 光源类型可能的值为 Directional Light(全局光)、Point Light(点光源)、Spot Light(聚光灯) 和 Area Light(区域光，只有在烘焙下才会体现)。

（1）Range，定义从对象中心发出的光线的行进距离[仅限点光源(Point)和聚光灯(Spot)]。

（2）Spot Angle，定义聚光灯锥形底部的角度(以度为单位)[仅限聚光灯(Spot)]。

2. Color

Color 使用拾色器来设置发光的颜色。

3. Mode

Mode 为指定光照模式，此模式用于确定是否以及如何"烘焙"光源。模式可能为 Realtime(实时)、Mixed(实时与烘焙之间)和 Baked(烘焙)。

4. Intensity

Intensity 设置光源的亮度。方向光(Direction)的默认值为 0.5。点光源(Point)、聚光灯（Spot）或面光源（Area）的默认值为 1。

（1）Indirect Multiplier。使用此值可改变间接光的强度。间接光是从一个对象弹射到另一个对象的光。Indirect Multiplier 定义由全局光照（GI）系统计算的散射光的亮度。如果将 Indirect Multiplier 设置为低于 1 的值，每次反弹都会使散射光变得更暗。大于 1 的值使光线在每次弹射之后更明亮。例如，将阴暗处的阴暗面（例如洞穴内部）变亮到能够清晰可见，这个非常有用。如果要使用实时全局光照，但是希望限制单一实时光源以便它只发出直射光，请将其 Indirect Multiplier 设置为 0。

（2）Shadow Type。决定此光源投射为生硬阴影、柔和阴影还是根本不投射阴影。请参阅有关阴影的文档以了解关于硬阴影以及软阴影的信息。

项目4 如何在Unity中编写脚本

4.1 项目表单

表 4-1 学习性工作任务单

学习场 K	如何在 Unity 中编写脚本					
学习情境 L	熟悉 C♯ 语言					
学习任务 M	运用 C♯ 语言编写脚本			学时		4 学时(180 min)
典型工作过程描述	创建项目—熟悉 C♯ 语言—运用 C♯ 语言编写脚本					
学习目标	1. 了解 C♯ 语言 2. 熟悉 C♯ 语言 3. 运用 C♯ 语言编写脚本					
任务描述	熟悉 C♯ 语言并运用 C♯ 语言编写脚本					
学时安排	资讯 20 min	计划 10 min	决策 10 min	实施 100 min	检查 20 min	评价 20 min
对学生的要求	1. 安装好软件 2. 课前做好预习 3. 熟悉 C♯ 语言 4. 运用 C♯ 语言编写脚本					
参考资料	1. 素材包 2. 微视频 3. PPT					

表 4-2 资讯单

学习场 K	如何在 Unity 中编写脚本		
学习情境 L	熟悉 C♯ 语言		
学习任务 M	运用 C♯ 语言编写脚本	学时	20 min
典型工作过程描述	创建项目—熟悉 C♯ 语言—运用 C♯ 语言编写脚本		
搜集资讯的方式	1. 教师讲解 2. 互联网查询 3. 同学交流		
资讯描述	查看教师提供的资料，获取信息，便于编写		
对学生的要求	1. 软件安装完成 2. 课前做好预习 3. 熟悉 C♯ 语言 4. 运用 C♯ 语言编写脚本		
参考资料	1. 素材包 2. 微视频 3. PPT		

表 4-3 计划单

学习场 K	如何在 Unity 中编写脚本			
学习情境 L	熟悉 C♯ 语言			
学习任务 M	运用 C♯ 语言编写脚本		学时	10 min
典型工作过程描述	创建项目—熟悉 C♯ 语言—运用 C♯ 语言编写脚本			
计划制订的方式	同学间分组讨论			
序号	工作步骤		注意事项	
1	创建一个新的项目			
2	熟悉 C♯ 语言			
3	运用 C♯ 语言编写脚本			
计划评价	班级		第_____组	组长签字
	教师签字		日期	
	评语：			

表 4-4　决策单

学习场 K	如何在 Unity 中编写脚本		
学习情境 L	熟悉 C♯ 语言		
学习任务 M	运用 C♯ 语言编写脚本	学时	10 min
典型工作过程描述	创建项目—熟悉 C♯ 语言—运用 C♯ 语言编写脚本		

			计划对比		
序号	计划的可行性	计划的经济性	计划的可操作性	计划的实施难度	综合评价
1					
2					
3					
4					
5					
6					
7					
8					
9					
10					

	班级		第＿＿＿组	组长签字	
	教师签字		日期		
决策评价	评语：				

表 4-5　实施单

学习场 K	如何在 Unity 中编写脚本		
学习情境 L	熟悉 C♯ 语言		
学习任务 M	运用 C♯ 语言编写脚本	学时	100 min
典型工作过程描述	创建项目—熟悉 C♯ 语言—运用 C♯ 语言编写脚本		

序号	实施步骤	注意事项
1	创建一个新的项目	新建项目
2	熟悉 C♯ 语言	C♯ 语言的用法
3	运用 C♯ 语言编写脚本	是否合理，是否正确

实施说明：
1. 启动 UnityHub 程序后，需要创建一个项目
2. 熟悉 C♯ 语言
3. 运用 C♯ 语言编写脚本

	班级		＿第＿＿＿组	组长签字	
	教师签字		日期		
实施评价	评语：				

表 4-6 检查单

学习场 K	如何在 Unity 中编写脚本			
学习情境 L	熟悉 C♯ 语言			
学习任务 M	运用 C♯ 语言编写脚本		学时	20 min
典型工作过程描述	创建项目—熟悉 C♯ 语言—运用 C♯ 语言编写脚本			
序号	检查项目	检查标准	学生自查	教师检查
1	资讯环节	获取相关信息情况		
2	计划环节	熟悉 C♯ 语言		
3	实施环节	运用 C♯ 语言编写脚本		
4	检查环节	各个环节逐一检查		
检查评价	班级		第____组	组长签字
	教师签字		日期	
	评语：			

表 4-7 评价单

学习场 K	如何在 Unity 中编写脚本			
学习情境 L	熟悉 C♯ 语言			
学习任务 M	运用 C♯ 语言编写脚本		学时	20 min
典型工作过程描述	创建项目—熟悉 C♯ 语言—运用 C♯ 语言编写脚本			
评价项目	评价子项目	学生自评	组内评价	教师评价
资讯环节	1. 听取教师讲解 2. 互联网查询情况 3. 同学交流情况			
计划环节	1. 查询资料情况 2. 熟悉 C♯ 语言			
实施环节	1. 学习态度 2. 对 C♯ 语言的掌握程度			
最终结果	综合情况			
评价	班级		第____组	组长签字
	教师签字		日期	
	评语：			

表 4-8　教学引导文设计单

学习场 K	如何在 Unity 中编写脚本	学习情境 L	熟悉 C♯语言	参照系	信息工程学院
		学习任务 M	运用 C♯语言编写脚本		
普适性工作过程　　典型工作过程	资讯	计划	决策	实施	检查	评价
分析 C♯语言	教师讲解	同学分组讨论	计划的可行性	了解 C♯语言	获取相关信息情况	评价学习态度
熟悉 C♯语言	自行掌握	熟悉 C♯语言	计划的经济性	设置操作方式	检查语句	评价学生的熟悉度
进行运用	运用 C♯语言编写脚本	设计操作	计划的实施难度	运用 C♯语言编写脚本	检查语句运用是否正确	软件熟练程度
保存项目文件	了解项目文件的格式	了解项目文件的格式	综合评价	保存项目文件	检查项目文件的格式	评价项目

表 4-9　教学反馈单(学生反馈)

学习场 K	如何在 Unity 中编写脚本			
学习情境 L	熟悉 C♯语言			
学习任务 M	运用 C♯语言编写脚本		学时	4 学时(180 min)
典型工作过程描述	创建项目—熟悉 C♯语言—运用 C♯语言编写脚本			
调查项目	序号	调查内容	理由描述	
	1	资讯环节		
	2	计划环节		
	3	实施环节		
	4	检查环节		
您对本次课程教学的改进意见：				
调查信息	被调查人姓名		调查日期	

表 4-10 分组单

学习场 K	如何在 Unity 中编写脚本			
学习情境 L	熟悉 C♯ 语言			
学习任务 M	运用 C♯ 语言编写脚本		学时	4 学时(180 min)
典型工作过程描述	创建项目—熟悉 C♯ 语言—运用 C♯ 语言编写脚本			
分组情况	组别	组长	组员	
	1			
	2			
	3			
	4			
	5			
	6			
	7			
	8			
分组说明				
班级		教师签字	日期	

表 4-11 教师实施计划单

学习场 K	如何在 Unity 中编写脚本					
学习情境 L	熟悉 C♯ 语言					
学习任务 M	运用 C♯ 语言编写脚本			学时	4 学时(180 min)	
典型工作过程描述	创建项目—熟悉 C♯ 语言—运用 C♯ 语言编写脚本					
序号	工作与学习步骤	学时	使用工具	地点	方式	备注
1	资讯情况	20 min	互联网			
2	计划情况	10 min	计算机			
3	决策情况	10 min	计算机			
4	实施情况	100 min	Unity			
5	检查情况	20 min	计算机			
6	评价情况	20 min	课程伴侣			
班级		教师签字			日期	

表 4-12 成绩报告单

_____班级如何在 Unity 中编写脚本学习场(课程)成绩报告单													
学习场 K	如何在 Unity 中编写脚本												
学习情境 M	熟悉 C♯ 语言												
典型工作过程描述	创建项目—熟悉 C♯ 语言—运用 C♯ 语言编写脚本								学时		4 学时(180 min)		
序号	姓名	第一个学习任务				第二个学习任务				第 N 个学习任务				总评
		自评 ×%	互评 ×%	教师评 ×%	合计	自评 ×%	互评 ×%	教师评 ×%	合计	自评 ×%	互评 ×%	教师评 ×%	合计	
1														

序号	姓名	第一个学习任务				第二个学习任务				第 N 个学习任务				总评
		自评 ×%	互评 ×%	教师评 ×%	合计	自评 ×%	互评 ×%	教师评 ×%	合计	自评 ×%	互评 ×%	教师评 ×%	合计	
2														
3														
4														
5														
6														
7														
8														
9														
10														
11														
12														
13														
14														
15														
16														
17														
18														
19														
20														
21														
22														
23														
24														
25														
26														
27														
28														
班级			教师签字							日期				

4.2　理论指导

如何在 **Unity**
中编写程序 **1**

C♯的基本语法和使用介绍如下。

1. C♯变量的定义和使用

(1)C♯变量的定义。在编程语言中，每一个数值都需要存储在某个特定的地方，称之为变量。变量由变量名称和数据类型构成。变量的数据类型决定了可以在其中存储哪种类型的数据。

如何在 **Unity**
中编写程序 **2**

在定义变量时，首先要确认在变量中存放的值的数据类型，然后确定变量的内容，最后根据C♯变量命名规则命名好变量名。

定义变量的语法如下：

数据类型 变量名；

例如定义一个存放整数的变量，可以定义成

int num；

在定义变量后给变量赋值，则可直接使用"＝"来连接将在变量中存放的值即可。

如何在 **Unity**
中编写程序 **3**

(2)常见的数据类型有如下几种：

①数字型的变量。在C♯中，数字型的变量主要包括整数(int)、单精度浮点数(float)和双精度浮点数(double)。float和double类型变量的区别：float有效位数是6位，占用4个字节的存储空间；而double有效位数是15位，占用8个字节的存储空间。默认情况下，小数用double来表示。如果需要使用float时，需要在末尾加上f。比如：float a＝1.23f；在上面代码中，定义了一个名为a的浮点型变量，初始值为1.23。

除了以上几种常用的数字变量类型，在C♯中还有其他类型的数字变量，见表4-13。其使用方法大同小异，这里不再赘述。

表4-13　C♯中的数字变量

关键字	说明	字节大小
bool	逻辑值(真/假)	1
sbyte	有符号8位整数	1
byte	无符号8位整数	1
short	有符号16位整数	2
ushort	无符号16位整数	2
int	有符号32位整数	4
uint	无符号32位整数	4
long	有符号64位整数	8
ulong	无符号64位整数	8
char	16位字符类型	2
float	32位单精度浮点类型	4
double	64位双精度浮点类型	8
decimal	128位高精度浮点类型	16

②文本型的变量。文本型的变量主要是 char 和 string。其中，char 类型变量用于保存单个字符的值；string 类型变量则用于保存字符串的值。比如：string a＝"hello";

上面的代码中，定义了一个名为 a 的文本型的变量，其类型为 string(字符串)，其初始值为 hello。也可以用 string 类型保存数字：string a＝"1";不过此时 a 中保存的数据字符串是"1"，并不是单纯的数字 1。

③布尔型的变量。在 C♯ 中，布尔型的变量是 bool，用于保存逻辑状态的变量，包含两个值：真(true)和假(flase)。比如：bool isPlay＝true;

上面代码定义了一个名为 isPlay 的布尔型变量，初始值为真。需要注意的是，C♯ 是大小写敏感，不能将大小写混淆。比如：Bool isPlay＝true;

该语句在程序中会报错，因为在 C♯ 中并没有 Bool 类型的变量，只有 bool 类型。

(3)此外，C♯ 还支持一种特殊的变量——常量。常量就是不能被改变的变量，常量只能定义在类属性级别，常量也必须是静态的，并且在定义时就初始化赋值，常量一旦被初始化后就不可以在改变，任何对常量改变语句都将引起编译器错误。因为常量定义时就要赋值所以常量类型只能声明给值类型。

在声明变量时，只需要在变量的前面加上关键字 const 或 readonly 即可把该变量指定为一个常量。比如：

const int constNum ＝ 2;

readonly int readonlyNum ＝ 3;

(4)常量及变量的命名规则。

①只能由字母、数字、@和下画线组成。

②不能以数字为开头。

③@符号只能放在首位。

④不能与系统变关键词重名。

⑤不能重名：C♯ 大小写敏感。

⑥中文变量名在语法上是成立的，但是最好不用。

(5)常量及变量的命名规范。

①用英文单词，不要用拼音。

②小驼峰命名法中第一个单词首字母不大写，后面每个单词的首字母大写如：myHeroDamage。

③见名知意。

2. C♯ 变量的赋值

赋值的语法有两种方式：一种是在定义变量的同时直接赋值；另一种是先定义变量然后赋值。它们的格式如下。

(1)在定义变量的同时赋值：

数据类型 变量名＝值;

(2)先定义变量然后赋值：

数据类型 变量名;

变量名＝值;

在定义变量时需要注意变量中的值要与变量的数据类型相兼容。另外，在为变量赋值时也可以一次为多个变量赋值。例如：

int a＝1，b＝2；

虽然一次为多个变量赋值方便了很多，但在实际编程中为了增强程序的可读性，建议在编程中每次声明一个变量并为一个变量赋值。

3. 表达式与运算符

表达式与运算符的作用是对数据或信息进行各种形式的运算处理，它们构成了程序代码的主体。

表达式由运算符和操作数组成。

运算符的分类和优先级见表 4-14。

表 4-14 运算符的分类和优先级

运算符分类		运算符优先级		
运算符类别	运算符	类别	计算顺序	运算符
基本算术运算	＋,－,＊,/,％	基本	高	x,y,f(x),a[x],x＋＋,x－－
递增、递减	＋＋,－－	一元		＋,－,!,~,＋＋x,－－x,(T)x
位移	＜＜,＞＞	乘除		＋,/,％
逻辑	&,|,∧,!,－,&&,||	加减		＋,－
赋值	＝,＋＝,－＝,＊＝,/＝,％＝,&＝,|＝,∧＝,＜＜＝,＞＞＝	位移		＜＜,＞＞
关系	＝＝,!＝,＜,＞,＜＝,＞＝	关系		＜,＞,＜＝,＞＝
字符串串联	＋	相等		＝＝,!＝
成员访问	.	逻辑 AND		&
索引	[]	逻辑 XOR		∧
转换	()	逻辑 OR		|
条件运算	?:	条件 AND		&&
		条件 OR		||
		条件		?:
		赋值	低	＝,＋＝,－＝,＊＝,/＝,％＝,&＝,|＝,∧＝,＜,＜＝,＞,＞＝

在 C♯ 中，需要了解以下几种主要的运算符。

(1)算术运算符。算术运算符是我们都很熟悉的基本算术运算符号，主要是加＋、减一、乘＊、除/、求余数％。

(2)赋值运算符。赋值运算符用于将一个数据赋予一个变量、属性或者引用。数据可以是常量、变量或者表达式。赋值运算符本身又分为简单赋值和复合赋值，见表 4-15。

表 4-15　赋值运算符

赋值运算符	表达式示例	含义
＝	x＝10	将 10 赋给变量 x
＋＝	x＋＝y	x＝x＋y
－＝	x－＝y	x＝x－y
＊＝	x＊＝y	x＝x＊y
/＝	x/＝y	x＝x/y
％＝	％＝y	x＝x％y
＞＞＝	x＞＞＝y	x＝x＞＞＝y
＜＜＝	x＜＜＝y	x＝x＜＜＝y
＆＝	x＆＝y	x＝x＆y
|＝	x|＝y	x＝x|y
∧＝	x∧＝y	x＝x∧y

（3）关系运算符。关系运算符用于比较两个值之间的关系，并在比较后返回一个布尔类型的运算结果。常用的关系运算符见表 4-16。

表 4-16　常用的关系运算符

关系运算符	作用说明
＝＝	等于
＜	小于
＜＝	小于或等于
＞	大于
＞＝	大于或等于
!＝	不等于

（4）条件运算符。条件运算符用于进行逻辑判断，并返回一个布尔类型的运算结果。常用的条件运算符见表 4-17。

表 4-17　常用的条件运算符

条件运算符	作用说明
＆＆	与
ll	或
!	非
?:	三目运算符

4. 流程控制

　　（1）if。

　　　　语法形式：

　　　　if(表达式)

```
    {
    语句;
    }
```

（2）if...else。

语法形式：

```
if(表达式)
{
语句块 1；
}
else
{
语句块 2；
}
```

（3）while。

语法形式：

```
while(表达式)
{
循环体
}
```

（4）do...while。

语法形式：

```
do
{
循环体语句；
}
while(条件表达式);
```

（5）for。

语法形式：

```
for(表达式 1；表达式 2);
{
循环体语句；
}
```

（6）foreach。

语法形式：

```
foreach(类型 变量名 in 集合对象)
{
语句体
}
```

5. 数组

数组是一个存储相同类型元素的固定大小的顺序集合。数组是用来存储数据的集合，通常认为数组是一个同一类型变量的集合。

声明数组。即在 C♯ 中声明一个数组，可以使用下面的语法：

如 datatype[ ] arrayName;

其中：datatype 用于指定被存储在数组中的元素的类型；[ ]指定数组的秩（维度），秩指定数组的大小；arrayName 指定数组的名称。

又如：double[ ] balance;

6. 类、对象、方法

在 C♯ 语言中创建的任何项目都有类的存在，通过类能很好地体现面向对象语言中封装、继承、多态的特性。类定义的语法形式并不复杂，请记住 class 关键字，它是定义类的关键字。

类定义的具体语法形式如下：

类的访问修饰符　修饰符　类名
{
　　类的成员
}

(1)类的访问修饰符：用于设定对类的访问限制，包括 public、internal 或者不写，用 internal 或者不写时，代表只能在当前项目中访问类；public 则代表可以在任何项目中访问类。

(2)修饰符：修饰符是对类本身特点的描述，包括 abstract、sealed 和 static。abstract 是抽象的意思，使用它，修饰符的类不能被实例化；sealed 修饰的类是密封类，不能被继承；static 修饰的类是静态类，不能被实例化。

(3)类名：类名用于描述类的功能，因此在定义类名时最好是具有实际意义，这样方便用户理解类中描述的内容。在同一个命名空间下类名必须是唯一的。

(4)类的成员：在类中能定义的元素，主要包括字段、属性、方法。

类（class）是最基础的 C♯ 类型。类是一个数据结构，将状态(字段)和操作(方法和其他函数成员)组合在一个单元中。类为动态创建的类实例（instance）提供了定义，实例也称为对象（object）。类支持继承（inheritance）和多态性（polymorphism），这是派生类（derived class）可用来扩展和专用化基类（base class）的机制。

使用类声明可以创建新的类。类声明以一个声明头开始，其组成方式如下：首先指定类的属性和修饰符；然后是类的名称；最后是基类(如有)以及该类实现的接口。声明头后面跟着类体，它由一组位于一对大括号"{"和"}"之间的成员声明组成。

当不再使用对象时，该对象占用的内存将自动收回。在 C♯ 中，没有必要也不可能显示释放分配给对象的内存。

7. 简述

(1)对象：现实世界中的实体(世间万物皆对象)。

(2)类：具有相似属性和方法的对象的集合。

（3）面向对象程序设计的特点：封装、继承、多态。

（4）对象的三要素：属性（对象是什么）、方法（对象能做什么）、事件（对象如何响应）。

8. Unity 脚本系统

脚本是所有游戏中必不可少的组成部分。

（1）Unity 脚本概述。尽管 Unity 对脚本使用标准 Mono 运行时的实现方案，但在从脚本访问引擎方面，仍然有自己的惯例和技术。

（2）Console 面板。Console 窗口菜单包含了用于打开日志文件、控制列表中显示的消息数量以及设置堆栈跟踪的选项。Console 工具栏包含用于控制显示的消息数量以及搜索和过滤消息的选项（图 4-1）。

图 4-1

①Clear：移除从代码中生成的所有消息，但会保留编译器错误。

②Collapse：仅显示重复消息的第一次出现。有时在每次帧更新时会生成的运行时错误（例如 null 引用），此选项在这种情况下非常有用。

③Clear On Play：每当进入播放模式时就会自动清空控制台。

④Clear on Build：在构建项目时清空控制台。

⑤Error Pause：只要从脚本中调用了 Debug. LogError 便暂停回放。

⑥［Attach-to-Player］：打开一个下拉菜单，其中包含的选项可用于连接到运行在远程设备上的开发版本以及在控制台中显示其播放器日志。

⑦Player Logging：如果控制台连接到远程开发版本，则此选项将对版本启用播放器日志记录。禁用此选项将暂停日志记录，但控制台仍将连接到目标版本，还会隐藏此下拉菜单中的其余选项。选择在 Player Logging 下方列出的任何版本可在 Console 窗口中显示其日志。

⑧Editor：如果控制台连接到远程开发版本，则选择此选项可显示来自本地 Unity Player 的日志，而不是来自远程版本的日志。

⑨＜Enter IP＞：打开 Enter Player IP 对话框，可以在此对话框中指定远程设备上开发版本的 IP 地址。

⑩Connect：单击对话框中的 Connect 按钮可连接到版本，并可将其添加到下拉菜单底部的开发版本列表中。

⑪［DEVELOPMENT BUILDS］：列出可用的开发版本。这包括自动检测到的版本以及使用 Enter IP 选项添加的版本。

⑫消息开关：显示控制台中的消息数量，单击可显示/隐藏消息。

⑬警告开关：显示控制台中的警告数量，单击可显示/隐藏警告。

⑭错误开关：显示控制台中的错误数量，单击可显示/隐藏错误。

（3）创建和使用脚本。游戏对象的行为由附加的组件控制。Unity 允许使用脚本来自行创建组件。使用脚本可以触发游戏事件，随时修改组件属性，并以所需的任何方式响应用户的输入。

Unity 本身支持 C# 编程语言。C#（发音为 C-sharp）是一种类似 Java 或 C++的行业标准语言。

C# # 创建脚本与大多数其他资源不同，脚本通常直接在 Unity 中创建。可以从 Project 面板左上方的 Create 菜单新建脚本，也可以通过从主菜单执行"Assets"→"Create"→"C# Script"命令来新建脚本（图 4-2）。

图 4-2

创建一个名为 FirstScripts 的脚本文件，双击 Unity 中的脚本资源时，将在文本编辑器中打开此脚本（图 4-3）。

其中的代码如下：

```
using System.Collections;
using System.Collections.Generic;
using UnityEngine;

public class FirstScript : MonoBehaviour
{
    // Start is called before the first frame update
    void Start()
    {

    }

    // Update is called once per frame
    void Update()
    {

    }
}
```

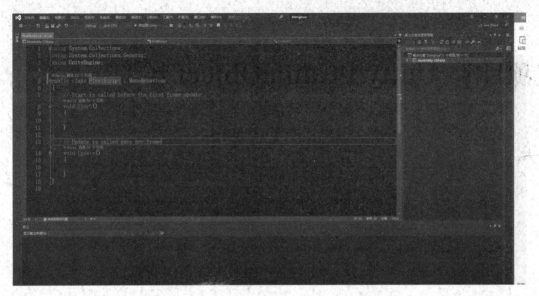

图 4-3

脚本中默认提供了 Start() 和 Update() 两个方法。这两个方法是 Unity 生命周期的一部分，都为 private 类型。其中 Start() 方法会在脚本激活的第一帧调用，而 Update() 方法在每一帧都会被调用，Update 函数是放置代码的地方，用于处理游戏对象的帧更新。这可能包括移动、触发动作和响应用户输入，基本上涉及游戏运行过程中随时间推移而需要处理的任何事项。

项目5　Unity中GameObject类

5.1　项目表单

表 5-1　学习性工作任务单

学习场 K	Unity 中 GameObject 类					
学习情境 L	熟悉 GameObject 类					
学习任务 M	运用 GameObject 类编写脚本				学时	4 学时(180 min)
典型工作过程描述	创建项目—熟悉 GameObject 类—运用 GameObject 类编写脚本					
学习目标	1. 了解 GameObject 类 2. 熟悉 GameObject 类 3. 运用 GameObject 类编写脚本					
任务描述	熟悉 GameObject 类并运用 GameObject 类编写脚本					
学时安排	资讯 20 min	计划 10 min	决策 10 min	实施 100 min	检查 20 min	评价 20 min
对学生的要求	1. 安装好软件 2. 课前做好预习 3. 熟悉 GameObject 类 4. 运用 GameObject 类编写脚本					
参考资料	1. 素材包 2. 微视频 3. PPT					

表 5-2 资讯单

学习场 K	Unity 中 GameObject 类		
学习情境 L	熟悉 GameObject 类		
学习任务 M	**运用 GameObject 类编写脚本**	**学时**	20 min
典型工作过程描述	创建项目—熟悉 GameObject 类—运用 GameObject 类编写脚本		
搜集资讯的方式	1. 教师讲解 2. 互联网查询 3. 同学交流		
资讯描述	查看教师提供的资料，获取信息，便于编写		
对学生的要求	1. 软件安装完成 2. 课前做好预习 3. 熟悉 GameObject 类 4. 运用 GameObject 类编写脚本		
参考资料	1. 素材包 2. 微视频 3. PPT		

表 5-3 计划单

学习场 K	Unity 中 GameObject 类			
学习情境 L	熟悉 GameObject 类			
学习任务 M	运用 GameObject 类编写脚本		学时	10 min
典型工作过程描述	创建项目—熟悉 GameObject 类—运用 GameObject 类编写脚本			
计划制订的方式	同学间分组讨论			
序号	工作步骤		注意事项	
1	创建一个新的项目			
2	熟悉 GameObject 类			
3	运用 GameObject 类编写脚本			
计划评价	**班级**		**第＿＿组**	**组长签字**	
	教师签字		**日期**		
	评语：				

表 5-4 决策单

学习场 K	Unity 中 GameObject 类			
学习情境 L	熟悉 GameObject 类语言			
学习任务 M	运用 GameObject 类编写脚本		学时	10 min
典型工作过程描述	创建项目—熟悉 GameObject 类—运用 GameObject 类编写脚本			
计划对比				
序号	计划的可行性	计划的经济性	计划的可操作性	计划的实施难度	综合评价
1					
2					
3					
4					
5					
6					
7					
8					
9					
10					
决策评价	班级		第____组	组长签字	
	教师签字		日期		
	评语：				

表 5-5 实施单

学习场 K	Unity 中 GameObject 类		
学习情境 L	熟悉 GameObject 类		
学习任务 M	运用 GameObject 类编写脚本	学时	100 min
典型工作过程描述	创建项目—熟悉 GameObject 类—运用 GameObject 类编写脚本		
序号	实施步骤	注意事项	
1	创建一个新的项目	新建项目	
2	熟悉 GameObject 类	GameObject 类的用法	
3	运用 GameObject 类编写脚本	是否合理，是否正确	

实施说明：
1. 启动 UnityHub 程序后，需要创建一个项目
2. 熟悉 GameObject 类
3. 运用 GameObject 类编写脚本

实施评价	班级		第____组	组长签字	
	教师签字		日期		
	评语：				

表 5-6 检查单

学习场 K	Unity 中 GameObject 类			
学习情境 L	熟悉 GameObject 类			
学习任务 M	运用 GameObject 类编写脚本		学时	20 min
典型工作过程描述	创建项目—熟悉 GameObject 类—运用 GameObject 类编写脚本			
序号	检查项目	检查标准	学生自查	教师检查
1	资讯环节	获取相关信息情况		
2	计划环节	熟悉 GameObject 类		
3	实施环节	运用 GameObject 类编写脚本		
4	检查环节	各个环节逐一检查		

检查评价	班级		第____组	组长签字	
	教师签字		日期		
	评语：				

表 5-7 评价单

学习场 K	Unity 中 GameObject 类			
学习情境 L	熟悉 GameObject 类			
学习任务 M	运用 GameObject 类编写脚本		学时	20 min
典型工作过程描述	创建项目—熟悉 GameObject 类—运用 GameObject 类编写脚本			
评价项目	评价子项目	学生自评	组内评价	教师评价
资讯环节	1. 听取教师讲解 2. 互联网查询情况 3. 同学交流情况			
计划环节	1. 查询资料情况 2. 熟悉 GameObject 类			
实施环节	1. 学习态度 2. 对 GameObject 类的掌握程度			
最终结果	综合情况			

评价	班级		第____组	组长签字	
	教师签字		日期		
	评语：				

表 5-8　教学引导文设计单

学习场 K	Unity 中 GameObject 类	学习情境 L	熟悉 GameObject 类	参照系	信息工程学院
		学习任务 M	运用 GameObject 类编写脚本		
普适性工作过程 / 典型工作过程	资讯	计划	决策	实施	检查	评价
分析 GameObject 类	教师讲解	同学分组讨论	计划的可行性	了解 GameObject 类	获取相关信息情况	评价学习态度
熟悉 GameObject 类	自行掌握	熟悉 GameObject 类	计划的经济性	设置操作方式	检查语句	评价学生的熟悉度
进行运用	运用 GameObject 类编写脚本	设计操作	计划的实施难度	运用 GameObject 类编写脚本	检查语句运用是否正确	软件熟练程度
保存项目文件	了解项目文件的格式	了解项目文件的格式	综合评价	保存项目文件	检查项目文件的格式	评价项目

表 5-9　教学反馈单(学生反馈)

学习场 K	Unity 中 GameObject 类			
学习情境 L	熟悉 GameObject 类			
学习任务 M	运用 GameObject 类编写脚本		学时	4 学时(180 min)
典型工作过程描述	创建项目—熟悉 GameObject 类—运用 GameObject 类编写脚本			
调查项目	序号	调查内容		理由描述
	1	资讯环节		
	2	计划环节		
	3	实施环节		
	4	检查环节		
您对本次课程教学的改进意见:				
调查信息	被调查人姓名		调查日期	

表 5-10 分组单

学习场 K	Unity 中 GameObject 类				
学习情境 L	熟悉 GameObject 类				
学习任务 M	运用 GameObject 类编写脚本			学时	4 学时(180 min)
典型工作过程描述	创建项目—熟悉 GameObject 类—运用 GameObject 类编写脚本				
分组情况	组别	组长		组员	
	1				
	2				
	3				
	4				
	5				
	6				
	7				
	8				
分组说明					
班级		教师签字		日期	

表 5-11 教师实施计划单

学习场 K	Unity 中 GameObject 类				
学习情境 L	熟悉 GameObject 类				
学习任务 M	运用 GameObject 类编写脚本			学时	4 学时(180 min)
典型工作过程描述	创建项目—熟悉 GameObject 类—运用 GameObject 类编写脚本				
序号	工作与学习步骤	学时	使用工具	地点	方式	备注
1	资讯情况	20 min	互联网			
2	计划情况	10 min	计算机			
3	决策情况	10 min	计算机			
4	实施情况	100 min	Unity			
5	检查情况	20 min	计算机			
6	评价情况	20 min	课程伴侣			
班级		教师签字		日期		

表 5-12 成绩报告单

___班级Unity 中 GameObject 类学习场(课程)成绩报告单															
学习场 K	Unity 中 GameObject 类														
学习情境 M	熟悉 GameObject 类														
典型工作过程描述	创建项目—熟悉 GameObject 类—运用 GameObject 类编写脚本									学时		4 学时(180 min)			
序号	姓名	第一个学习任务				第二个学习任务				第 N 个学习任务				总评
		自评 ×%	互评 ×%	教师评 ×%	合计	自评 ×%	互评 ×%	教师评 ×%	合计	自评 ×%	互评 ×%	教师评 ×%	合计	
1														

续表

序号	姓名	第一个学习任务				第二个学习任务				第 N 个学习任务				总评
		自评 ×%	互评 ×%	教师评 ×%	合计	自评 ×%	互评 ×%	教师评 ×%	合计	自评 ×%	互评 ×%	教师评 ×%	合计	
2														
3														
4														
5														
6														
7														
8														
9														
10														
11														
12														
13														
14														
15														
16														
17														
18														
19														
20														
21														
22														
23														
24														
25														
26														
27														
28														
班级			教师签字							日期				

5.2　理论指导

GameObject 类是 Unity 场景中所有实体的基类，其中的很多变量已被删除。

在 Unity 开发中使用最多的对象，莫过于 GameObject 类。以下列举了 GameObject 类中的一些方法案例。

1. 创建对象

创建对象有很多的方式：

(1)通过 GameObject 菜单栏中创建。

(2)通过代码的创建。

(3)从资源中拖拉进场景。

GameObject 类 1　　　**GameObject 类 2**

2. 脚本中常用属性和方法

(1)activeInHierarchy。定义 GameObject 在 Scene 中是否处于活动状态(图 5-1)。

GameObject.activeInHierarchy

public bool **activeInHierarchy** ;

描述

定义 GameObject 在 Scene 中是否处于活动状态。

这可使您知道 GameObject 在游戏中是否处于活动状态。如果启用了其 GameObject.activeSelf 属性，以及其所有父项的这一属性，则该 GameObject 处于活动状态。

图 5-1

activeInHierarchy 所表示的是物体在场景中实际的状态(图 5-2)。

```
// Start is called before the first frame update
void Start()
{
    Debug.Log(gameObject.activeInHierarchy);
}
```

图 5-2

(2)activeSelf。显示此 GameObject 的本地活动状态(只读)如图 5-3 所示。

图 5-3

activeself 只能显示物体本地的激活状态，如果想知道实际的激活状态，请使用 activeInHierarchy(图 5-4)。

图 5-4

(3)isStatic。仅限 Editor 的 API，指定游戏对象是否为静态(图 5-5)。

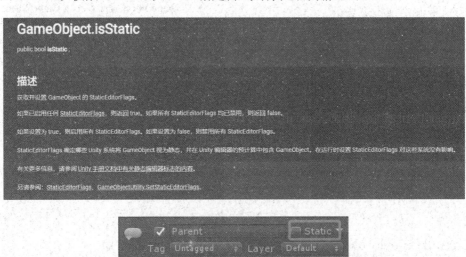

图 5-5

3. 构造函数

构造函数可以用来创建游戏对象实例(图 5-6)。

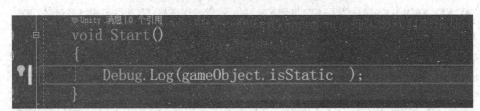

图 5-6

（1）Find。按 name 查找 GameObject，然后返回。此函数仅作为返回活动 GameObject（图 5-7）。

图 5-7

（2）FindGameObjectsWithTag。返回标记为 tag 的活动 GameObject 的列表。如果未找到 GameObject，则返回空数组（图 5-8）。

图 5-8

（3）FindWithTag。返回一个标记为 tag 的活动 GameObject。如果未找到 GameObject，则返回 null（图 5-9）。

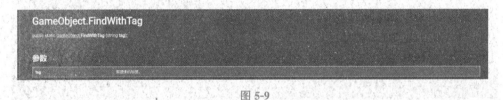

图 5-9

（4）GetComponent。如果游戏对象附加了类型为 type 的组件，则将其返回，否则返回 null（图 5-10）。

图 5-10

（5）SetActive。根据给定的值 true 或 false，激活或停用 GameObject。GameObject 可能因为父项未处于活动状态而处于非活动状态。在这种情况下，调用 SetActive 不会激活它，而是仅设置此 GameObject 的本地状态，该状态可使用 GameObject.activeSelf 加以检查。当所有父项均处于活动状态时，Unity 便可以使用此状态（图 5-11）。

图 5-11

（6）Destroy。删除 GameObject、组件或资源。实际的对象销毁操作始终延迟到当前 Update 循环结束、但始终在渲染前完成（图 5-12）。

图 5-12

（7）Instantiate。克隆 original 对象并返回克隆对象。该函数以与 Editor 中的"Duplicate"命令类似的方式创建对象的副本。如果要克隆的是 GameObject，也可以选择指定其位置和旋转（否则，默认为原始 GameObject 的位置和旋转）。如果克隆的是 Component，也将克隆其附加到的 GameObject，此时也可指定可选的位置和旋转。克隆 GameObject 或 Component 时，也将克隆所有子对象和组件，它们的属性设置与原始对象相同。克隆时将传递 GameObject 的激活状态，因此，如果原始对象处于非激活状态，则克隆对象也将创建为非激活状态（图 5-13）。

图 5-13

项目6　Unity中Transform类

6.1　项目表单

表 6-1　学习性工作任务单

学习场 K	Unity 中 Transform 类					
学习情境 L	熟悉 Transform 类					
学习任务 M	运用 Transform 类编写脚本			学时		4 学时(180 min)
典型工作过程描述	创建项目—熟悉 Transform 类—运用 Transform 类编写脚本					
学习目标	1. 了解 Transform 类 2. 熟悉 Transform 类 3. 运用 Transform 类编写脚本					
任务描述	熟悉 Transform 类并运用 Transform 类编写脚本					
学时安排	资讯 20 min	计划 10 min	决策 10 min	实施 100 min	检查 20 min	评价 20 min
对学生的要求	1. 安装好软件 2. 课前做好预习 3. 熟悉 Transform 类 4. 运用 Transform 类编写脚本					
参考资料	1. 素材包 2. 微视频 3. PPT					

表 6-2　资讯单

学习场 K	Unity 中 Transform 类		
学习情境 L	熟悉 Transform 类		
学习任务 M	运用 Transform 类编写脚本	学时	20 min
典型工作过程描述	创建项目—熟悉 Transform 类—运用 Transform 类编写脚本		
搜集资讯的方式	1. 教师讲解 2. 互联网查询 3. 同学交流		
资讯描述	查看教师提供的资料，获取信息，便于编写		
对学生的要求	1. 软件安装完成 2. 课前做好预习 3. 熟悉 Transform 类 4. 运用 Transform 类编写脚本		
参考资料	1. 素材包 2. 微视频 3. PPT		

表 6-3　计划单

学习场 K	Unity 中 Transform 类		
学习情境 L	熟悉 Transform 类		
学习任务 M	运用 Transform 类编写脚本	学时	10 min
典型工作过程描述	创建项目—熟悉 Transform 类—运用 Transform 类编写脚本		
计划制订的方式	同学间分组讨论		
序号	工作步骤	注意事项	
1	创建一个新的项目		
2	熟悉 Transform 类		
3	运用 Transform 类编写脚本		
计划评价	班级		第____组	组长签字
	教师签字		日期	
	评语：			

表 6-4　决策单

学习场 K	Unity 中 Transform 类			
学习情境 L	熟悉 Transform 类语言			
学习任务 M	运用 Transform 类编写脚本		学时	10 min
典型工作过程描述	创建项目—熟悉 Transform 类—运用 Transform 编写脚本			

计划对比

序号	计划的可行性	计划的经济性	计划的可操作性	计划的实施难度	综合评价
1					
2					
3					
4					
5					
6					
7					
8					
9					
10					

决策评价	班级		第____组	组长签字	
	教师签字		日期		
	评语：				

表 6-5　实施单

学习场 K	Unity 中 Transform 类		
学习情境 L	熟悉 Transform 类		
学习任务 M	运用 Transform 类编写脚本	学时	100 min
典型工作过程描述	创建项目—熟悉 Transform 类—运用 Transform 类编写脚本		

序号	实施步骤	注意事项
1	创建一个新的项目	新建项目
2	熟悉 Transform 类	Transform 类的用法
3	运用 Transform 类编写脚本	是否合理，是否正确

实施说明：

1. 启动 UnityHub 程序后，需要创建一个项目
2. 熟悉 Transform 类
3. 运用 Transform 类编写脚本

实施评价	班级		第____组	组长签字	
	教师签字		日期		
	评语：				

表 6-6　检查单

学习场 K	Unity 中 Transform 类			
学习情境 L	熟悉 Transform 类			
学习任务 M	运用 Transform 类编写脚本		学时	20 min
典型工作过程描述	创建项目—熟悉 Transform 类—运用 Transform 类编写脚本			
序号	检查项目	检查标准	学生自查	教师检查
1	资讯环节	获取相关信息情况		
2	计划环节	熟悉 Transform 类		
3	实施环节	运用 Transform 类编写脚本		
4	检查环节	各个环节逐一检查		
检查评价	班级		第＿＿＿组	组长签字	
	教师签字		日期		
	评语：				

表 6-7　评价单

学习场 K	Unity 中 Transform 类			
学习情境 L	熟悉 Transform 类			
学习任务 M	运用 Transform 类编写脚本		学时	20 min
典型工作过程描述	创建项目—熟悉 Transform 类—运用 Transform 类编写脚本			
评价项目	评价子项目	学生自评	组内评价	教师评价
资讯环节	1. 听取教师讲解 2. 互联网查询情况 3. 同学交流情况			
计划环节	1. 查询资料情况 2. 熟悉 Transform 类			
实施环节	1. 学习态度 2. 对 Transform 类的掌握程度			
最终结果	综合情况			
评价	班级		第＿＿＿组	组长签字	
	教师签字		日期		
	评语：				

表 6-8 教学引导文设计单

学习场 K	Unity 中 Transform 类	学习情境 L	熟悉 Transform 类	参照系	信息工程学院
		学习任务 M	运用 Transform 类编写脚本		
典型工作过程＼普适性工作过程	资讯	计划	决策	实施	检查	评价
分析 Transform 类	教师讲解	同学分组讨论	计划的可行性	了解 Transform 类	获取相关信息情况	评价学习态度
熟悉 Transform 类	自行掌握	熟悉 Transform 类	计划的经济性	设置操作方式	检查语句	评价学生的熟悉度
进行运用	运用 Transform 类编写脚本	设计操作	计划的实施难度	运用 Transform 类编写脚本	检查语句运用是否正确	软件熟练程度
保存项目文件	了解项目文件的格式	了解项目文件的格式	综合评价	保存项目文件	检查项目文件的格式	评价项目

表 6-9 教学反馈单(学生反馈)

学习场 K	Unity 中 Transform 类			
学习情境 L	熟悉 Transform 类			
学习任务 M	运用 Transform 类编写脚本		学时	4 学时(180 min)
典型工作过程描述	创建项目—熟悉 Transform 类—运用 Transform 类编写脚本			
调查项目	序号	调查内容		理由描述
	1	资讯环节		
	2	计划环节		
	3	实施环节		
	4	检查环节		

您对本次课程教学的改进意见：

调查信息	被调查人姓名		调查日期	

表 6-10　分组单

学习场 K	Unity 中 Transform 类				
学习情境 L	熟悉 Transform 类				
学习任务 M	运用 Transform 类编写脚本		学时	4 学时(180 min)	
典型工作过程描述	创建项目—熟悉 Transform 类—运用 Transform 类编写脚本				
分组情况	组别	组长	组员		
	1				
	2				
	3				
	4				
	5				
	6				
	7				
	8				
分组说明					
班级		教师签字		日期	

表 6-11　教师实施计划单

学习场 K	Unity 中 Transform 类					
学习情境 L	熟悉 Transform 类					
学习任务 M	运用 Transform 类编写脚本		学时	4 学时(180 min)		
典型工作过程描述	创建项目—熟悉 Transform 类—运用 Transform 类编写脚本					
序号	工作与学习步骤	学时	使用工具	地点	方式	备注
1	资讯情况	20 min	互联网			
2	计划情况	10 min	计算机			
3	决策情况	10 min	计算机			
4	实施情况	100 min	Unity			
5	检查情况	20 min	计算机			
6	评价情况	20 min	课程伴侣			
班级		教师签字		日期		

表 6-12 成绩报告单

序号	姓名	第一个学习任务				第二个学习任务				第 N 个学习任务				总评
		自评 ×%	互评 ×%	教师评 ×%	合计	自评 ×%	互评 ×%	教师评 ×%	合计	自评 ×%	互评 ×%	教师评 ×%	合计	
1														
2														
3														
4														
5														
6														
7														
8														
9														
10														
11														
12														
13														
14														
15														
16														
17														
18														
19														
20														
21														
22														
23														
24														
25														
26														
27														
28														

_____班级 Unity 中 Transform 类学习场(课程)成绩报告单

学习场 K	Unity 中 Transform 类		
学习情境 M	熟悉 Transform 类		
典型工作过程描述	创建项目—熟悉 Transform 类—运用 Transform 类编写脚本	学时	4 学时(180 min)

班级		教师签字		日期	

Transform 类 1

6.2　理论指导

1. Transform 组件概述

Transform 组件确定每个对象在场景中的 Position(位置)、Rotation(旋转)和 Scale(缩放)属性的值。每个游戏对象都有一个变换组件。

属性组件：Position、Rotation、Scale(图 6-1)。

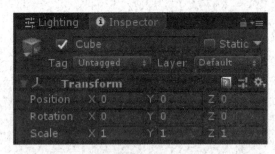

图 6-1

Transform 类 2

Transform 类 3

2. Transform 类(对象的位置、旋转和缩放)

场景中的每个对象都可以有一个变换。它用于存储和操作对象的位置、旋转和缩放。每个变换都可以有一个父级，能够分层应用位置、旋转和缩放(图 6-2)。

图 6-2

(1)Position。Transform 组件下的 Position 属性是一个相对于父对象的局部坐标，如果其没有父对象，其属性框里显示的自然而然也就是世界坐标。如果有父对象，那么在 Unity 编辑器中显示的值是以这个 Transform 以上一级父对象为基准的坐标。而 Transform. position 表示当前 Component 的全局坐标(图 6-3)。

图 6-3

(2)Rotation 与 eulerAngles。Unity 中的旋转分为四元数 Quaternion(x，y，z，w)和欧拉角 eulerAngles(x，y，z)。

使用 Transform. rotation. z 获取到的是一个 cos，sin 的值，而不是面板中的值。而使用 Transform. eulerAngles. z 是一个 0～360 的值，当正向旋转时可以获取到面板上对应的数值，但反向旋转时数值就会从 360 一直往下减少。

rotation：一个四元数，用于存储变换在世界空间中的旋转(Quaternion 四元数)(图 6-4)。

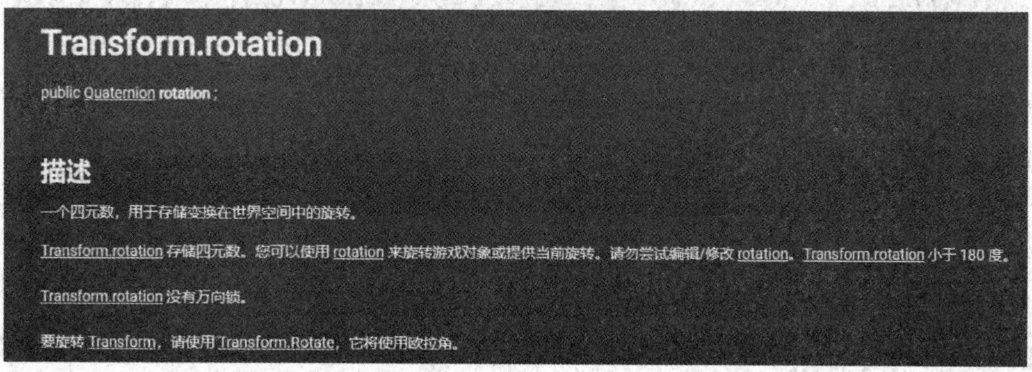

Transform.rotation

public Quaternion rotation ;

描述

一个四元数，用于存储变换在世界空间中的旋转。

Transform.rotation 存储四元数。您可以使用 rotation 来旋转游戏对象或提供当前旋转。请勿尝试编辑/修改 rotation。Transform.rotation 小于 180 度。

Transform.rotation 没有万向锁。

要旋转 Transform，请使用 Transform.Rotate，它将使用欧拉角。

图 6-4

(3)eulerAngles。以欧拉角表示的旋转(以度为单位)(Vector3)(图 6-5)。

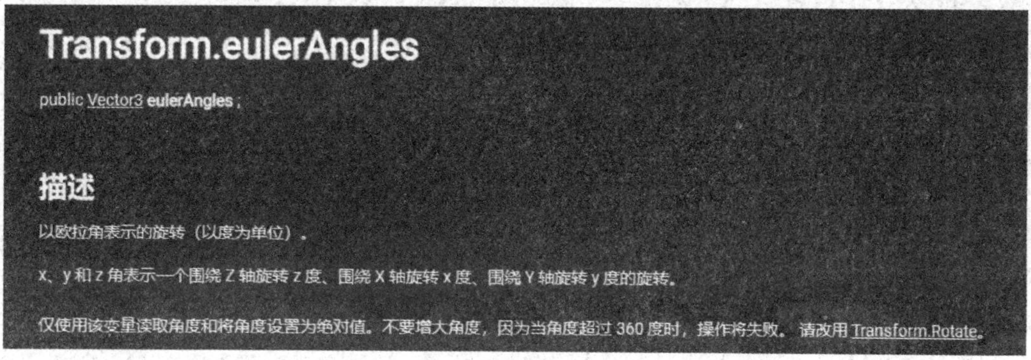

Transform.eulerAngles

public Vector3 eulerAngles ;

描述

以欧拉角表示的旋转（以度为单位）。

x、y 和 z 角表示一个围绕 Z 轴旋转 z 度、围绕 X 轴旋转 x 度、围绕 Y 轴旋转 y 度的旋转。

仅使用该变量读取角度和将角度设置为绝对值。不要增大角度，因为当角度超过 360 度时，操作将失败。请改用 Transform.Rotate。

图 6-5

3. 本地与世界

在 Unity 中，父子化是一个非常重要的概念。当一个游戏对象是另一个游戏对象的父物体时，其子游戏对象会随着它移动、旋转和缩放，就像胳膊属于身体，当旋转身体时，胳膊也会跟着旋转一样。但子物体移动、旋转和缩放，父物体不会随其改变。任何物体都可以有多个子物体，但只能有一个父物体(图 6-6)。

localEulerAngles：以欧拉角表示的相对于父变换旋转的旋转(以度为单位)。

localPosition：相对于父变换的变换位置。

localRotation：相对于父变换旋转的变换旋转。

localScale：相对于父对象的变换缩放。

lossyScale 对象的全局缩放(只读)。

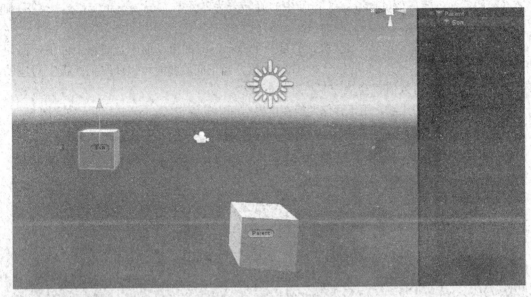

图 6-6

（1）设置物体父物体。在脚本中公开属性，可以将其暴露在所属物体的 Inspector（检视视图）面板中（图 6-7）。脚本代码如下：

```
using System.Collections;
using System.Collections.Generic;
using UnityEngine;
public class NewBehaviourScript : MonoBehaviour
{
    public int num= 3 ;
    // Start is called before the first frame update
    void Start()
    {

    }
}
```

（值以 Inspector 面板上的数据为准。）

Parent：变换的父级（图 6-8）。脚本代码如下：

```
using System.Collections;
using System.Collections.Generic;
using UnityEngine;
public class NewBehaviourScript : MonoBehaviour
{
    public Transform parent Tran ;
```

```
// Start is called before the first frame update
void Start()
{
    Transform. parent= parentTran ;
}
}
```

图 6-7

图 6-8

(2)Translate(位移设置)(图 6-9)。

```
public void Translate(float x, float y, float z);
public void Translate(float x, float y, float z, [DefaultValue("Space.Self")] Space relativeTo);
public void Translate(Vector3 translation);
public void Translate(Vector3 translation, [DefaultValue("Space.Self")] Space relativeTo);
public void Translate(float x, float y, float z, Transform relativeTo);
public void Translate(Vector3 translation, Transform relativeTo);
```

图 6-9

根据 Translation 的方向和距离移动变换(图 6-10)。

图 6-10

项目7　Prefab预制体

7.1　项目表单

表 7-1　学习性工作任务单

学习场 K	Prefab 预制体					
学习情境 L	了解 Prefab 预制体					
学习任务 M	编写预制体的脚本及预制体的实例化			学时		4 学时(180 min)
典型工作过程描述	创建项目—了解 Prefab 预制体—编写预制体的脚本及预制体的实例化					
学习目标	1. 了解 Prefab 预制体 2. 熟悉 Prefab 预制体 3. 编写预制体的脚本及预制体的实例化					
任务描述	熟悉 Prefab 预制体并编写预制体的脚本及预制体的实例化					
学时安排	资讯 20 min	计划 10 min	决策 10 min	实施 100 min	检查 20 min	评价 20 min
对学生的要求	1. 安装好软件 2. 课前做好预习 3. 熟悉 Prefab 预制体 4. 运用编写预制体的脚本及预制体的实例化					
参考资料	1. 素材包 2. 微视频 3. PPT					

表 7-2　资讯单

学习场 K	Prefab 预制体		
学习情境 L	了解 Prefab 预制体		
学习任务 M	编写预制体的脚本及预制体的实例化	学时	20 min
典型工作过程描述	创建项目—了解 Prefab 预制体—编写预制体的脚本及预制体的实例化		
搜集资讯的方式	1. 教师讲解 2. 互联网查询 3. 同学交流		
资讯描述	查看教师提供的资料，获取信息，便于操作		
对学生的要求	1. 软件安装完成 2. 课前做好预习 3. 熟悉 Prefab 预制体 4. 编写预制体的脚本及预制体的实例化		
参考资料	1. 素材包 2. 微视频 3. PPT		

表 7-3　计划单

学习场 K	Prefab 预制体			
学习情境 L	了解 Prefab 预制体			
学习任务 M	编写预制体的脚本及预制体的实例化		学时	10 min
典型工作过程描述	创建项目—了解 Prefab 预制体—编写预制体的脚本及预制体的实例化			
计划制订的方式	同学间分组讨论			
序号	工作步骤		注意事项	
1	创建一个新的项目			
2	熟悉 Prefab 预制体			
3	编写预制体的脚本及预制体的实例化			
计划评价	班级		第＿＿＿组	组长签字
	教师签字		日期	
	评语：			

表7-4 决策单

学习场 K	Prefab 预制体		
学习情境 L	了解 Prefab 预制体		
学习任务 M	编写预制体的脚本及预制体的实例化	学时	10 min
典型工作过程描述	创建项目—了解 Prefab 预制体—编写预制体的脚本及预制体的实例化		
计划对比					
序号	计划的可行性	计划的经济性	计划的可操作性	计划的实施难度	综合评价
1					
2					
3					
4					
5					
6					
7					
8					
9					
10					

决策评价	班级		第＿＿组	组长签字	
	教师签字		日期		
	评语：				

表7-5 实施单

学习场 K	Prefab 预制体		
学习情境 L	了解 Prefab 预制体		
学习任务 M	编写预制体的脚本及预制体的实例化	学时	100 min
典型工作过程描述	创建项目—了解 Prefab 预制体—编写预制体的脚本及预制体的实例化		
序号	实施步骤		注意事项
1	创建一个新的项目		新建项目
2	了解 Prefab 预制体		Prefab 预制体
3	编写预制体的脚本及预制体的实例化		是否合理，是否正确

实施说明：

1. 启动 UnityHub 程序后，需要创建一个项目
2. 熟悉 Prefab 预制体
3. 编写预制体的脚本及预制体的实例化

实施评价	班级		第＿＿组	组长签字	
	教师签字		日期		
	评语：				

表 7-6 检查单

学习场 K	Prefab 预制体			
学习情境 L	了解 Prefab 预制体			
学习任务 M	编写预制体的脚本及预制体的实例化		学时	20 min
典型工作过程描述	创建项目—了解 Prefab 预制体—编写预制体的脚本及预制体的实例化			
序号	检查项目	检查标准	学生自查	教师检查
1	资讯环节	获取相关信息情况		
2	计划环节	熟悉 Prefab 预制体		
3	实施环节	编写预制体的脚本及预制体的实例化		
4	检查环节	各个环节逐一检查		
检查评价	班级		第＿＿＿组	组长签字
	教师签字		日期	
	评语：			

表 7-7 评价单

学习场 K	Prefab 预制体			
学习情境 L	了解 Prefab 预制体			
学习任务 M	编写预制体的脚本及预制体的实例化		学时	20 min
典型工作过程描述	创建项目—了解 Prefab 预制体—编写预制体的脚本及预制体的实例化			
评价项目	评价子项目	学生自评	组内评价	教师评价
资讯环节	1. 听取教师讲解 2. 互联网查询情况 3. 同学交流情况			
计划环节	1. 查询资料情况 2. 熟悉 Prefab 预制体			
实施环节	1. 学习态度 2. 对 Prefab 预制体的掌握程度			
最终结果	综合情况			
评价	班级		第＿＿＿组	组长签字
	教师签字		日期	
	评语：			

表7-8　教学引导文设计单

学习场 K	Prefab 预制体	学习情境 L	熟悉 Prefab 预制体	参照系		信息工程学院
		学习任务 M	编写预制体的脚本及预制体的实例化			
典型工作过程　＼　普适性工作过程	资讯	计划	决策	实施	检查	评价
分析 Prefab 预制体	教师讲解	同学分组讨论	计划的可行性	了解 Prefab 预制体	获取相关信息情况	评价学习态度
熟悉 Prefab 预制体	自行掌握	熟悉 Prefab 预制体	计划的经济性	设置操作方式	检查语句	评价学生的熟悉度
进行运用	编写预制体的脚本及预制体的实例化	设计操作	计划的实施难度	运用 Prefab 预制体	检查语句运用是否正确	软件熟练程度
保存项目文件	了解项目文件的格式	了解项目文件的格式	综合评价	保存项目文件	检查项目文件的格式	评价项目

表7-9　教学反馈单(学生反馈)

学习场 K	Prefab 预制体			
学习情境 L	了解 Prefab 预制体			
学习任务 M	编写预制体的脚本及预制体的实例化		学时	4 学时(180 min)
典型工作过程描述	创建项目—了解 Prefab 预制体—编写预制体的脚本及预制体的实例化			
调查项目	序号	调查内容		理由描述
	1	资讯环节		
	2	计划环节		
	3	实施环节		
	4	检查环节		
您对本次课程教学的改进意见：				
调查信息	被调查人姓名		调查日期	

表 7-10　分组单

学习场 K	Prefab 预制体				
学习情境 L	了解 Prefab 预制体				
学习任务 M	编写预制体的脚本及预制体的实例化			学时	4 学时(180 min)
典型工作过程描述	创建项目—了解 Prefab 预制体—编写预制体的脚本及预制体的实例化				
分组情况	组别	组长	组员		
	1				
	2				
	3				
	4				
	5				
	6				
	7				
	8				
分组说明					
班级		教师签字		日期	

表 7-11　教师实施计划单

学习场 K	Prefab 预制体					
学习情境 L	了解 Prefab 预制体					
学习任务 M	编写预制体的脚本及预制体的实例化			学时	4 学时(180 min)	
典型工作过程描述	创建项目—了解 Prefab 预制体—编写预制体的脚本及预制体的实例化					
序号	工作与学习步骤	学时	使用工具	地点	方式	备注
1	资讯情况	20 min	互联网			
2	计划情况	10 min	计算机			
3	决策情况	10 min	计算机			
4	实施情况	100 min	Unity			
5	检查情况	20 min	计算机			
6	评价情况	20 min	课程伴侣			
班级		教师签字		日期		

表 7-12 成绩报告单

班级Prefab 预制体学习场(课程)成绩报告单													
学习场 K	Prefab 预制体												
学习情境 M	了解 Prefab 预制体												
典型工作过程描述	创建项目—了解 Prefab 预制体—编写预制体的脚本及预制体的实例化							学时		4 学时(180 min)			
序号	姓名	第一个学习任务				第二个学习任务				第 N 个学习任务				总评
		自评 ×%	互评 ×%	教师评 ×%	合计	自评 ×%	互评 ×%	教师评 ×%	合计	自评 ×%	互评 ×%	教师评 ×%	合计	
1														
2														
3														
4														
5														
6														
7														
8														
9														
10														
11														
12														
13														
14														
15														
16														
17														
18														
19														
20														
21														
22														
23														
24														
25														
26														
27														
28														
班级		教师签字							日期					

7.2　理论指导

通过 Hierarchy 面板下的 Create 菜单可以手动地创建一个 GameObject，当要在程序里面动态地创建一个游戏物体的时候，可以新建一个 GameObject。Unity 提供一种被称为 Prefab 的预置对象，它是以文件的形式保存在硬盘上的一个 GameObject，它里面可能包含了各种设置、组件及一些脚本。Prefab 允许在不同的 Scene，甚至 Project 中使用同一个对象，例如实现了一个子弹，通过打包成 Prefab，可以在另外一个游戏里面直接使用它。

Prefab 预制体

1. 创建预制体

(1)直接将 Cube 拖拽到 Project 面板中 assets 里建立一个预制体(图 7-1)

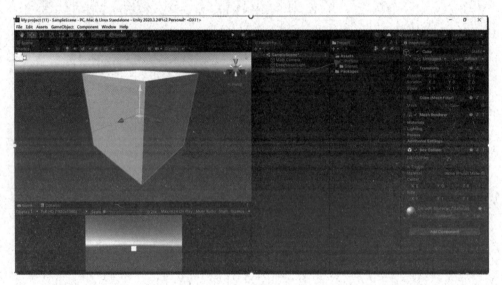

图 7-1

(2)可以观察到 Cube 变成蓝色，Project 面板中出现了一个 Cube(图 7-2)。

(3)选中场景中的 Cube，可以看到在 Inspector 面板中多了这样一栏，单击 Select 可以定位到 Project 中预制体的模板"母体"，单击 Revert 就是还原当前预制体，Apply 就是确定当前预制体(图 7-3)。

2. 实例化预制体(Instantiate)

该函数以与 Editor 中的"Duplicate"命令类似的方式创建对象的副本。如果要克隆的是 GameObject，也可以选择指定其位置和旋转(否则默认为原始 GameObject 的位置和旋转)。如果克隆的是 Component，也将克隆其附加到的 GameObject，此时也可指定可选的位置和旋转。克隆 GameObject 或 Component 时，也将克隆所有子对象和组件，它们的属性设置与原始对象相同。在默认情况下，新对象的父对象将为 null，因此它与原始对象不"同级"。但是，可以使用重载方法设置父对象。如果指定了父对象但未指定位置和旋转，则使用原始对象的位置和旋转作为克隆对象的本地位置和旋转；如果 instanti-ateInWorldSpace 参数为 true，则使用原始对象的世界位置和旋转。如果指定了位置和旋转，则使用它们作为该对象在世界空间中的位置和旋转。克隆时将传递 GameObject 的激活状态，因此，如果原始对象处于非激活状态，则克隆对象也将创建为非激活状态（图7-4）。

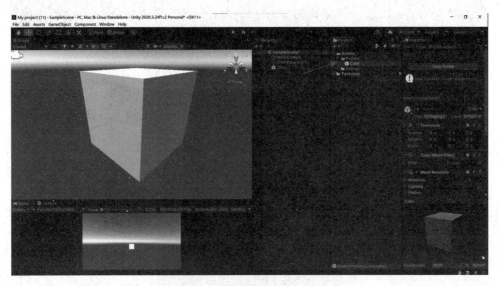

图 7-2

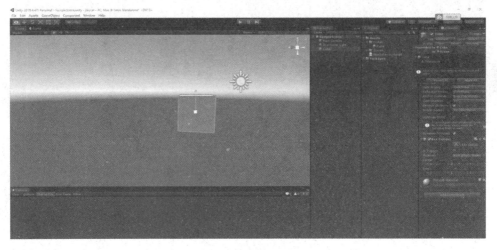

图 7-3

3. 预制体的实例化

直接将预制体拖拽到 Hierarchy 面板中，可以发现 Hierarchy 面板中预制体变蓝了（图7-5）。

图 7-4

图 7-5

项目8　Collider碰撞器

8.1　项目表单

表 8-1　学习性工作任务单

学习场 K	Collider 碰撞器					
学习情境 L	了解 Collider 碰撞器					
学习任务 M	根据所学创建一个完整项目			学时		4 学时(180 min)
典型工作过程描述	创建项目—了解 Collider 碰撞器—根据所学创建一个完整项目					
学习目标	1. 了解 Collider 碰撞器 2. 熟悉 Collider 碰撞器 3. 根据所学创建一个完整项目					
任务描述	熟悉 Collider 碰撞器 Prefab 预制体并根据所学创建一个完整项目					
学时安排	资讯 20 min	计划 10 min	决策 10 min	实施 100 min	检查 20 min	评价 20 min
对学生的要求	1. 安装好软件 2. 课前做好预习 3. 熟悉 Collider 碰撞器 4. 根据所学创建一个完整项目					
参考资料	1. 素材包 2. 微视频 3. PPT					

表 8-2 资讯单

学习场 K	Collider 碰撞器		
学习情境 L	了解 Collider 碰撞器		
学习任务 M	根据所学创建一个完整项目	学时	20 min
典型工作过程描述	创建项目—了解 Collider 碰撞器—根据所学创建一个完整项目		
搜集资讯的方式	1. 教师讲解 2. 互联网查询 3. 同学交流		
资讯描述	查看教师提供的资料，获取信息，便于创建		
对学生的要求	1. 软件安装完成 2. 课前做好预习 3. 熟悉 Collider 碰撞器 4. 根据所学创建一个完整项目		
参考资料	1. 素材包 2. 微视频 3. PPT		

表 8-3 计划单

学习场 K	Collider 碰撞器		
学习情境 L	了解 Collider 碰撞器		
学习任务 M	根据所学创建一个完整项目	学时	10 min
典型工作过程描述	创建项目—了解 Collider 碰撞器—根据所学创建一个完整项目		
计划制订的方式	同学间分组讨论		
序号	工作步骤	注意事项	
1	创建一个新的项目		
2	熟悉 Collider 碰撞器		
3	根据所学创建一个完整项目		
计划评价	班级	第____组	组长签字
	教师签字	日期	
	评语：		

表 8-4 决策单

学习场 K	Collider 碰撞器		
学习情境 L	了解 Collider 碰撞器		
学习任务 M	根据所学创建一个完整项目	学时	10 min
典型工作过程描述	创建项目—了解 Collider 碰撞器—根据所学创建一个完整项目		

计划对比					
序号	计划的可行性	计划的经济性	计划的可操作性	计划的实施难度	综合评价
1					
2					
3					
4					
5					
6					
7					
8					
9					
10					

决策评价	班级		第＿＿＿组	组长签字	
	教师签字		日期		
	评语：				

表 8-5 实施单

学习场 K	Collider 碰撞器		
学习情境 L	了解 Collider 碰撞器		
学习任务 M	根据所学创建一个完整项目	学时	100 min
典型工作过程描述	创建项目—了解 Collider 碰撞器—根据所学创建一个完整项目		

序号	实施步骤	注意事项
1	创建一个新的项目	新建项目
2	了解 Collider 碰撞器	Collider 碰撞器
3	根据所学创建一个完整项目	是否合理，是否正确

实施说明：
1. 启动 UnityHub 程序后，需要创建一个项目
2. 熟悉 Collider 碰撞器
3. 根据所学创建一个完整项目

实施评价	班级		第＿＿＿组	组长签字	
	教师签字		日期		
	评语：				

表 8-6　检查单

学习场 K	Collider 碰撞器			
学习情境 L	了解 Collider 碰撞器			
学习任务 M	根据所学创建一个完整项目		学时	20 min
典型工作过程描述	创建项目—了解 Collider 碰撞器—根据所学创建一个完整项目			
序号	检查项目	检查标准	学生自查	教师检查
1	资讯环节	获取相关信息情况		
2	计划环节	熟悉 Collider 碰撞器		
3	实施环节	根据所学创建一个完整项目		
4	检查环节	各个环节逐一检查		
检查评价	班级		第＿＿＿组	组长签字	
	教师签字		日期		
	评语：				

表 8-7　评价单

学习场 K	Collider 碰撞器			
学习情境 L	了解 Collider 碰撞器			
学习任务 M	根据所学创建一个完整项目		学时	20 min
典型工作过程描述	创建项目—了解 Collider 碰撞器—根据所学创建一个完整项目			
评价项目	评价子项目	学生自评	组内评价	教师评价
资讯环节	1. 听取教师讲解 2. 互联网查询情况 3. 同学交流情况			
计划环节	1. 查询资料情况 2. 熟 Collider 碰撞器			
实施环节	1. 学习态度 2. 对 Collider 碰撞器的掌握程度			
最终结果	综合情况			
评价	班级		第＿＿＿组	组长签字	
	教师签字		日期		
	评语：				

表 8-8　教学引导文设计单

学习场 K	Collider 碰撞器	学习情境 L	了解 Collider 碰撞器	参照系	信息工程学院
		学习任务 M	根据所学创建一个完整项目		
典型工 作过程　　普适性工 作过程	资讯	计划	决策	实施	检查	评价
分析 Collider 碰撞器	教师讲解	同学分组讨论	计划的可行性	了解 Collider 碰撞器	获取相关 信息情况	评价学习 态度
熟悉 Collider 碰撞器	自行掌握	熟悉 Collider 碰撞器	计划的经济性	设置操作方式	检查组件	评价学生 的熟悉度
进行运用	根据所学创建 一个完整项目	设计操作	计划的 实施难度	运用所学 所有知识	检查语句 运用是否正确	软件 熟练程度
保存项目文件	了解项目 文件的格式	了解项目 文件的格式	综合评价	保存项目文件	检查项目 文件的格式	评价项目

表 8-9　教学反馈单(学生反馈)

学习场 K	Collider 碰撞器			
学习情境 L	了解 Collider 碰撞器			
学习任务 M	根据所学创建一个完整项目		学时	4 学时(180 min)
典型工作过程描述	创建项目—了解 Collider 碰撞器—根据所学创建一个完整项目			
调查项目	序号	调查内容		理由描述
	1	资讯环节		
	2	计划环节		
	3	实施环节		
	4	检查环节		

您对本次课程教学的改进意见:

调查信息	被调查人姓名		调查日期	

表 8-10　分组单

学习场 K	Collider 碰撞器			
学习情境 L	了解 Collider 碰撞器			
学习任务 M	根据所学创建一个完整项目		学时	4 学时（180 min）
典型工作过程描述	创建项目—了解 Collider 碰撞器—根据所学创建一个完整项目			
分组情况	组别	组长	组员	
	1			
	2			
	3			
	4			
	5			
	6			
	7			
	8			
分组说明				
班级		教师签字		日期

表 8-11　教师实施计划单

学习场 K	Collider 碰撞器					
学习情境 L	了解 Collider 碰撞器					
学习任务 M	根据所学创建一个完整项目			学时	4 学时（180 min）	
典型工作过程描述	创建项目—了解 Collider 碰撞器—根据所学创建一个完整项目					
序号	工作与学习步骤	学时	使用工具	地点	方式	备注
1	资讯情况	20 min	互联网			
2	计划情况	10 min	计算机			
3	决策情况	10 min	计算机			
4	实施情况	100 min	Unity			
5	检查情况	20 min	计算机			
6	评价情况	20 min	课程伴侣			
班级		教师签字			日期	

表 8-12　成绩报告单

班级Collider 碰撞器学习场(课程)成绩报告单													
学习场 K	Collider 碰撞器												
学习情境 M	了解 Collider 碰撞器												
典型工作过程描述	创建项目—了解 Collider 碰撞器—根据所学创建一个完整项目								学时		4 学时(180 min)			
序号	姓名	第一个学习任务				第二个学习任务				第 N 个学习任务				总评
		自评×%	互评×%	教师评×%	合计	自评×%	互评×%	教师评×%	合计	自评×%	互评×%	教师评×%	合计	
1														
2														
3														
4														
5														
6														
7														
8														
9														
10														
11														
12														
13														
14														
15														
16														
17														
18														
19														
20														
21														
22														
23														
24														
25														
26														
27														
28														
班级		教师签字							日期					

8.2 理论指导

碰撞体（Collider）组件定义对象的形状以便用于物理碰撞。碰撞体是不可见的，其形状不需要与对象的网格完全相同，事实上，粗略近似方法通常更有效，在游戏运行过程中难以察觉。

Unity 提供了 4 种不同的 Collider 组件，分别如下：

（1）Box Collider：立方体状的 Collider，Cube 对象默认挂载该 Collider；

（2）Capsule Collider：胶囊状的 Collider，Capsule 对象默认挂载该 Collider；

Collider
碰撞器

（3）Sphere Collider：球状的 Collider，Sphere 对象默认挂载该 Collider；

（4）Mesh Collider：根据 Mesh 确定形状的 Collider，Plane 对象默认挂载该 Collider。

1. 碰撞简单讲解

最简单（并且也是处理器开销最低）的碰撞体是所谓的原始碰撞体类型。在 3D 中，这些碰撞体为盒形碰撞体、球形碰撞体和胶囊碰撞体。在 2D 中，可以使用 2D 盒形碰撞体和 2D 圆形碰撞体。可以将任意数量的上述碰撞体添加到单个对象以创建复合碰撞体。

可将碰撞体添加到没有刚体组件的对象，从而创建场景的地板、墙壁和其他静止元素，这些被称为静态碰撞体。在通常情况下，不应通过更改变换位置来重新定位静态碰撞体，因为这会极大地影响物理引擎的性能。具有刚体的对象上的碰撞体称为动态碰撞体。静态碰撞体可与动态碰撞体相互作用，但由于没有刚体，因此不会通过移动来响应碰撞（图 8-1）。

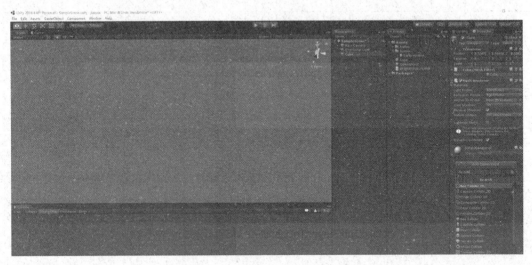

图 8-1

2. 主要脚本介绍

（1）OnCollisionEnter。当该碰撞体/刚体已开始接触另一个刚体/碰撞体时，自动调用 OnCollisionEnter（图 8-2）。

图 8-2

(2)OnCollisionExit。当该碰撞体/刚体已停止接触另一个刚体/碰撞体时，自动调用 OnCollisionExit(图 8-3)。

图 8-3

(3)OnCollisionStay。对应正在接触刚体/碰撞体的每一个碰撞体/刚体，每帧自动调用一次 OnCollisionStay(图 8-4)。

图 8-4

(4)OnTriggerEnter。当 Collider other 事件进入该触发器时自动调用 OnTriggerEnter。此消息被发送到触发器 Collider 和触发器 Collider 所属的 Rigidbody(如果有)，以及接触该触发器的 Rigidbody 或 Collider(如果没有 Rigidbody)(图 8-5)。

图 8-5

(5)OnTriggerExit。当 Collider other 已停止接触该触发器时自动调用 OnTriggerExit(图 8-6)。

图 8-6

（6）OnTriggerStay。对于正在接触该触发器的每个其他 Collider，绝大多数帧自动调用 OnTriggerStay。此函数位于物理计时器上，因此它不必运行每个帧(图 8-7)。

图 8-7

项目9　Rigidbody刚体

9.1　项目表单

表 9-1　学习性工作任务单

学习场 K	Rigidbody 刚体				
学习情境 L	了解 Rigidbody 刚体				
学习任务 M	根据刚体和碰撞器结合创建一个完整项目		学时	4 学时(180 min)	
典型工作过程描述	创建项目—了解 Rigidbody 刚体—根据刚体和碰撞器结合创建一个完整项目				
学习目标	1. 了解 Rigidbody 刚体 2. 熟悉 Rigidbody 刚体 3. 根据刚体和碰撞器结合创建一个完整项目				
任务描述	熟悉 Rigidbody 刚体并根据刚体和碰撞器结合创建一个完整项目				
学时安排	资讯 20 min	计划 10 min	决策 10 min	实施 100 min	检查 20 min	评价 20 min
对学生的要求	1. 安装好软件 2. 课前做好预习 3. 熟悉 Rigidbody 刚体 4. 根据刚体和碰撞器结合创建一个完整项目				
参考资料	1. 素材包 2. 微视频 3. PPT				

表 9-2 资讯单

学习场 K	Rigidbody 刚体		
学习情境 L	了解 Rigidbody 刚体		
学习任务 M	根据刚体和碰撞器结合创建一个完整项目	学时	20 min
典型工作过程描述	创建项目—了解 Rigidbody 刚体—根据刚体和碰撞器结合创建一个完整项目		
搜集资讯的方式	1. 教师讲解 2. 互联网查询 3. 同学交流		
资讯描述	查看教师提供的资料，获取信息，便于创建		
对学生的要求	1. 软件安装完成 2. 课前做好预习 3. 熟悉 Rigidbody 刚体 4. 根据刚体和碰撞器结合创建一个完整项目		
参考资料	1. 素材包 2. 微视频 3. PPT		

表 9-3 计划单

学习场 K	Rigidbody 刚体			
学习情境 L	了解 Rigidbody 刚体			
学习任务 M	根据刚体和碰撞器结合创建一个完整项目		学时	10 min
典型工作过程描述	创建项目—了解 Rigidbody 刚体—根据刚体和碰撞器结合创建一个完整项目			
计划制订的方式	同学间分组讨论			
序号	工作步骤		注意事项	
1	创建一个新的项目			
2	熟悉 Rigidbody 刚体			
3	根据刚体和碰撞器结合创建一个完整项目			
计划评价	班级		第___组	组长签字
	教师签字		日期	
	评语：			

表 9-4　决策单

学习场 K	Rigidbody 刚体			
学习情境 L	了解 Rigidbody 刚体			
学习任务 M	根据刚体和碰撞器结合创建一个完整项目		学时	10 min
典型工作过程描述	创建项目—了解 Rigidbody 刚体—根据刚体和碰撞器结合创建一个完整项目			
计划对比				
序号	计划的可行性	计划的经济性	计划的可操作性	计划的实施难度	综合评价
1					
2					
3					
4					
5					
6					
7					
8					
9					
10					

决策评价	班级		第　　组		组长签字	
	教师签字		日期			
	评语：					

表 9-5　实施单

学习场 K	Rigidbody 刚体		
学习情境 L	了解 Rigidbody 刚体		
学习任务 M	根据刚体和碰撞器结合创建一个完整项目	学时	100 min
典型工作过程描述	创建项目—了解 Rigidbody 刚体—根据刚体和碰撞器结合创建一个完整项目		
序号	实施步骤	注意事项	
1	创建一个新的项目	新建项目	
2	了解 Rigidbody 刚体	Rigidbody 刚体	
3	根据刚体和碰撞器结合创建一个完整项目	是否合理，是否正确	

实施说明：

1. 启动 UnityHub 程序后，需要创建一个项目
2. 熟悉 Rigidbody 刚体
3. 根据刚体和碰撞器结合创建一个完整项目

实施评价	班级		第　　组		组长签字	
	教师签字		日期			
	评语：					

表 9-6 检查单

学习场 K	Rigidbody 刚体			
学习情境 L	了解 Rigidbody 刚体			
学习任务 M	根据刚体和碰撞器结合创建一个完整项目		学时	20 min
典型工作过程描述	创建项目—了解 Rigidbody 刚体—根据刚体和碰撞器结合创建一个完整项目			
序号	检查项目	检查标准	学生自查	教师检查
1	资讯环节	获取相关信息情况		
2	计划环节	熟悉 Rigidbody 刚体		
3	实施环节	根据刚体和碰撞器 结合创建一个完整项目		
4	检查环节	各个环节逐一检查		
检查评价	班级		第____组	组长签字	
	教师签字		日期		
	评语：				

表 9-7 评价单

学习场 K	Rigidbody 刚体			
学习情境 L	了解 Rigidbody 刚体			
学习任务 M	根据刚体和碰撞器结合创建一个完整项目		学时	20 min
典型工作过程描述	创建项目—了解 Rigidbody 刚体—根据刚体和碰撞器结合创建一个完整项目			
评价项目	评价子项目	学生自评	组内评价	教师评价
资讯环节	1. 听取教师讲解 2. 互联网查询情况 3. 同学交流情况			
计划环节	1. 查询资料情况 2. 熟悉 Rigidbody 刚体			
实施环节	1. 学习态度 2. 对 Rigidbody 刚体的掌握程度			
最终结果	综合情况			
评价	班级		第____组	组长签字	
	教师签字		日期		
	评语：				

表 9-8 教学引导文设计单

学习场 K	Rigidbody 刚体	学习情境 L	了解 Rigidbody 刚体	参照系	信息工程学院
		学习任务 M	根据刚体和碰撞器结合创建一个完整项目		
典型工作过程 ／ 普适性工作过程	资讯	计划	决策	实施	检查	评价
分析 Rigidbody 刚体	教师讲解	同学分组讨论	计划的可行性	了解 Rigidbody 刚体	获取相关信息情况	评价学习态度
熟悉 Rigidbody 刚体	自行掌握	熟悉 Rigidbody 刚体	计划的经济性	设置操作方式	检查组件	评价学生的熟悉度
进行运用	根据刚体和碰撞器结合创建一个完整项目	设计操作	计划的实施难度	运用所学所有知识	检查语句运用是否正确	软件熟练程度
保存项目文件	了解项目文件的格式	了解项目文件的格式	综合评价	保存项目文件	检查项目文件的格式	评价项目

表 9-9 教学反馈单(学生反馈)

学习场 K	Rigidbody 刚体			
学习情境 L	了解 Rigidbody 刚体			
学习任务 M	根据刚体和碰撞器结合创建一个完整项目		学时	4 学时(180 min)
典型工作过程描述	创建项目—了解 Rigidbody 刚体—根据刚体和碰撞器结合创建一个完整项目			
调查项目	序号	调查内容		理由描述
	1	资讯环节		
	2	计划环节		
	3	实施环节		
	4	检查环节		

您对本次课程教学的改进意见：

调查信息	被调查人姓名		调查日期	

表 9-10 分组单

学习场 K	Rigidbody 刚体			
学习情境 L	了解 Rigidbody 刚体			
学习任务 M	根据刚体和碰撞器结合创建一个完整项目		学时	4 学时(180 min)
典型工作过程描述	创建项目—了解 Rigidbody 刚体—根据刚体和碰撞器结合创建一个完整项目			
	组别	组长		组员
	1			
	2			
	3			
	4			
分组情况	5			
	6			
	7			
	8			
分组说明				
班级		教师签字		日期

表 9-11 教师实施计划单

学习场 K	Rigidbody 刚体					
学习情境 L	了解 Rigidbody 刚体					
学习任务 M	根据刚体和碰撞器结合创建一个完整项目			学时	4 学时(180 min)	
典型工作过程描述	创建项目—了解 Rigidbody 刚体—根据刚体和碰撞器结合创建一个完整项目					
序号	工作与学习步骤	学时	使用工具	地点	方式	备注
1	资讯情况	20 min	互联网			
2	计划情况	10 min	计算机			
3	决策情况	10 min	计算机			
4	实施情况	100 min	Unity			
5	检查情况	20 min	计算机			
6	评价情况	20 min	课程伴侣			
班级		教师签字		日期		

表 9-12　成绩报告单

_____班级Rigidbody 刚体学习场(课程)成绩报告单													
学习场 K	Rigidbody 刚体												
学习情境 M	了解 Rigidbody 刚体												
典型工作过程描述	创建项目—了解 Rigidbody 刚体—根据刚体和碰撞器结合创建一个完整项目							学时		4 学时(180 min)			
序号	姓名	第一个学习任务				第二个学习任务				第 N 个学习任务				总评
		自评 ×%	互评 ×%	教师评 ×%	合计	自评 ×%	互评 ×%	教师评 ×%	合计	自评 ×%	互评 ×%	教师评 ×%	合计	
1														
2														
3														
4														
5														
6														
7														
8														
9														
10														
11														
12														
13														
14														
15														
16														
17														
18														
19														
20														
21														
22														
23														
24														
25														
26														
27														
28														
班级		教师签字							日期					

9.2 理论指导

刚体（Rigidbody）使游戏对象（GameObject）的行为方式受物理控制。刚体可以接受力和扭矩，使对象以逼真的方式移动。任何游戏对象都必须包含受重力影响的刚体，行为方式基于施加的作用力（通过脚本），或通过 NVIDIA PhysX 物理引擎与其他对象交互（图 9-1）。

Rigidbody 刚体

1. 属性讲解

（1）Mass：为对象的质量（单位：kg）。

（2）Drag：为空气阻力。根据力移动对象时，影响对象的空气阻力大小。

（3）Angular Drag：为角阻力。根据扭矩旋转对象时，影响对象的空气阻力大小。

（4）Use Gravity：为开启重力。如果启用此属性，则对象将受重力影响（默认勾选）。

（5）Is Kinematic：为开启动力学模式。如果启用此选项，则对象将不会被物理引擎驱动，只能通过变换（Transform）对其进行操作。

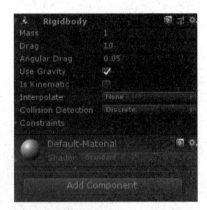

图 9-1

（6）Interpolate：为插值。仅当在刚体运动中看到急动时才尝试使用提供的选项之一。

①Interpolate：为内插值。根据前一帧的变换来平滑变换。

②Extrapolate：为外插值。根据下一帧的估计变换来平滑变换。

（7）None：为不应用插值。

（8）Collision Detection：为碰撞检测模式。用于防止快速移动的对象穿过其他对象而不检测碰撞。

①Discrete：为对场景中的所有其他碰撞体使用离散碰撞检测，其他碰撞体在测试碰撞时会使用离散碰撞检测。用于正常碰撞（这是默认值）。

②Continuous：为对动态碰撞体（具有刚体）使用离散碰撞检测，并对静态碰撞体（没有刚体）使用基于扫掠的连续碰撞检测。设置为连续动态（Continuous Dynamic）的刚体将在测试与该刚体的碰撞时使用连续碰撞检测。其他刚体将使用离散碰撞检测。用于连续动态（Continuous Dynamic）检测需要碰撞的对象（此属性对物理性能有很大影响，如果没有快速对象的碰撞问题，请将其保留为 Discrete 设置）。

③Continuous Dynamic：为对设置为连续（Continuous）和连续动态（Continuous Dynamic）碰撞的游戏对象使用基于扫掠的连续碰撞检测。还将对静态碰撞体（没有刚体）使用连续碰撞检测。对于所有其他碰撞体，使用离散碰撞检测。其用于快速移动的对象。

④Continuous Speculative：为对刚体和碰撞体使用推测性连续碰撞检测。这也是可以设置运动物体的唯一 CCD 模式。该方法通常比基于扫掠的连续碰撞检测的成本更低。

（9）Constraints：为对刚体运动的限制。

①Freeze Position：为位置冻结有选择地停止刚体沿世界 X、Y 和 Z 轴的移动。

②Freeze Rotation：为角度冻结有选择地停止刚体围绕局部 X、Y 和 Z 轴旋转。

2. AddForce（向 Rigidbody 添加力）

沿 force 矢量的方向连续施加力，可以指定 ForceMode/mode/，以将力的类型更改为 Acceleration、Impulse 或 Velocity Change。力只能应用于处于活动状态的刚体。如果 GameObject 处于非活动状态，则 AddForce 没有效果。默认情况下，一旦施加力（Vector3. zero 力除外），刚体的状态就会被设置为唤醒。

3. ForceMode 属性

ForceMode 有以下 4 个属性（图 9-2）。

（1）Force：给 Rigidbody 添加一个可持续的力，受 Mass 影响。

（2）Acceleration：给 Rigidbody 添加一个可持续的加速度，忽略 Mass 影响。

（3）Impluse：立即给 Rigidbody 添加一个冲力，受 Mass 影响。

（4）VelocityChange：立即给 Rigidbody 添加速度，忽略 Mass 影响。

Properties

Force	Add a continuous force to the rigidbody, using its mass.
Acceleration	Add a continuous acceleration to the rigidbody, ignoring its mass.
Impulse	Add an instant force impulse to the rigidbody, using its mass.
VelocityChange	Add an instant velocity change to the rigidbody, ignoring its mass.

图 9-2

项目10　Input输入系统

10.1　项目表单

表 10-1　学习性工作任务单

学习场 K	Input 输入系统					
学习情境 L	了解 Input 输入系统					
学习任务 M	根据 Input 输入系统创建一个完整项目			学时		4 学时(180 min)
典型工作过程描述	创建项目—了解 Input 输入系统—根据 Input 输入系统创建一个完整项目					
学习目标	1. 了解 Input 输入系统 2. 熟悉 Input 输入系统 3. 根据 Input 输入系统创建一个完整项目					
任务描述	熟悉 Input 输入系统并根据 Input 输入系统创建一个完整项目					
学时安排	资讯 20 min	计划 10 min	决策 10 min	实施 100 min	检查 20 min	评价 20 min
对学生的要求	1. 安装好软件 2. 课前做好预习 3. 熟悉 Input 输入系统 4. 根据 Input 输入系统创建一个完整项目					
参考资料	1. 素材包 2. 微视频 3. PPT					

表 10-2　资　讯　单

学习场 K	Input 输入系统		
学习情境 L	了解 Input 输入系统		
学习任务 M	根据 Input 输入系统创建一个完整项目	学时	20 min
典型工作过程描述	创建项目—了解 Input 输入系统体—根据 Input 输入系统创建一个完整项目		
搜集资讯的方式	1. 教师讲解 2. 互联网查询 3. 同学交流		
资讯描述	查看教师提供的资料，获取信息，便于创建		
对学生的要求	1. 软件安装完成 2. 课前做好预习 3. 熟悉 Input 输入系统 4. 根据 Input 输入系统创建一个完整项目		
参考资料	1. 素材包 2. 微视频 3. PPT		

表 10-3　计　划　单

学习场 K	Input 输入系统		
学习情境 L	了解 Input 输入系统		
学习任务 M	根据 Input 输入系统创建一个完整项目	学时	10 min
典型工作过程描述	创建项目—了解 Input 输入系统—根据 Input 输入系统创建一个完整项目		
计划制订的方式	同学间分组讨论		
序号	工作步骤	注意事项	
1	创建一个新的项目		
2	熟悉 Input 输入系统		
3	根据 Input 输入系统创建一个完整项目		
计划评价	班级		第____组	组长签字	
	教师签字		日期		
	评语：				

表 10-4　决策单

学习场 K	Input 输入系统		
学习情境 L	了解 Input 输入系统		
学习任务 M	根据 Input 输入系统创建一个完整项目	学时	10 min
典型工作过程描述	创建项目—了解 Input 输入系统—根据 Input 输入系统创建一个完整项目		

计划对比					
序号	计划的可行性	计划的经济性	计划的可操作性	计划的实施难度	综合评价
1					
2					
3					
4					
5					
6					
7					
8					
9					
10					

决策评价	班级		第　　组	组长签字	
	教师签字		日期		
	评语：				

表 10-5　实施单

学习场 K	Input 输入系统		
学习情境 L	了解 Input 输入系统		
学习任务 M	根据 Input 输入系统创建一个完整项目	学时	100 min
典型工作过程描述	创建项目—了解 Input 输入系统—根据 Input 输入系统创建一个完整项目		

序号	实施步骤	注意事项
1	创建一个新的项目	新建项目
2	了解 Input 输入系统	Input 输入系统
3	根据 Input 输入系统创建一个完整项目	是否合理，是否正确

实施说明：

1. 启动 UnityHub 程序后，需要创建一个项目
2. 熟悉 Input 输入系统
3. 根据 Input 输入系统创建一个完整项目

实施评价	班级		第　　组	组长签字	
	教师签字		日期		
	评语：				

表 10-6 检查单

学习场 K	Input 输入系统			
学习情境 L	了解 Input 输入系统			
学习任务 M	根据 Input 输入系统创建一个完整项目		学时	20 min
典型工作过程描述	创建项目—了解 Input 输入系统—根据 Input 输入系统创建一个完整项目			
序号	检查项目	检查标准	学生自查	教师检查
1	资讯环节	获取相关信息情况		
2	计划环节	熟悉 Input 输入系统		
3	实施环节	根据 Input 输入系统创建一个完整项目		
4	检查环节	各个环节逐一检查		
检查评价	班级		第　　组	组长签字	
	教师签字		日期		
	评语：				

表 10-7 评价单

学习场 K	Input 输入系统			
学习情境 L	了解 Input 输入系统			
学习任务 M	根据 Input 输入系统创建一个完整项目		学时	20 min
典型工作过程描述	创建项目—了解 Input 输入系统—根据 Input 输入系统创建一个完整项目			
评价项目	评价子项目	学生自评	组内评价	教师评价
资讯环节	1. 听取教师讲解 2. 互联网查询情况 3. 同学交流情况			
计划环节	1. 查询资料情况 2. 熟悉 Input 输入系统			
实施环节	1. 学习态度 2. 对 Input 输入系统的掌握程度			
最终结果	综合情况			
评价	班级		第　　组	组长签字	
	教师签字		日期		
	评语：				

表 10-8　教学引导文设计单

学习场 K	Input 输入系统	学习情境 L	了解 Input 输入系统	参照系	信息工程学院
		学习任务 M	根据 Input 输入系统创建一个完整项目		
普适性工作过程／典型工作过程	资讯	计划	决策	实施	检查	评价
分析 Input 输入系统	教师讲解	同学分组讨论	计划的可行性	了解 Input 输入系统	获取相关信息情况	评价学习态度
熟悉 Input 输入系统	自行掌握	熟悉 Input 输入系统	计划的经济性	设置操作方式	检查组件	评价学生的熟悉度
进行运用	根据 Input 输入系统创建一个完整项目	设计操作	计划的实施难度	运用所学所有知识	检查语句运用是否正确	软件熟练程度
保存项目文件	了解项目文件的格式	了解项目文件的格式	综合评价	保存项目文件	检查项目文件的格式	评价项目

表 10-9　教学反馈单(学生反馈)

学习场 K	Input 输入系统			
学习情境 L	了解 Input 输入系统			
学习任务 M	根据 Input 输入系统创建一个完整项目		学时	4 学时(180 min)
典型工作过程描述	创建项目—了解 Input 输入系统—根据 Input 输入系统创建一个完整项目			
调查项目	序号	调查内容		理由描述
	1	资讯环节		
	2	计划环节		
	3	实施环节		
	4	检查环节		

您对本次课程教学的改进意见：

调查信息	被调查人姓名		调查日期	

表 10-10 分组单

学习场 K	Input 输入系统				
学习情境 L	了解 Input 输入系统				
学习任务 M	根据 Input 输入系统创建一个完整项目		学时	4 学时(180 min)	
典型工作过程描述	创建项目—了解 Input 输入系统—根据 Input 输入系统创建一个完整项目				
分组情况	组别	组长	组员		
	1				
	2				
	3				
	4				
	5				
	6				
	7				
	8				
分组说明					
班级		教师签字		日期	

表 10-11 教师实施计划单

学习场 K	Input 输入系统				
学习情境 L	了解 Input 输入系统				
学习任务 M	根据 Input 输入系统创建一个完整项目		学时	4 学时(180 min)	
典型工作过程描述	创建项目—了解 Input 输入系统—根据 Input 输入系统创建一个完整项目				
序号	工作与学习步骤	学时	使用工具	地点	方式	备注
1	资讯情况	20 min	互联网			
2	计划情况	10 min	计算机			
3	决策情况	10 min	计算机			
4	实施情况	100 min	Unity			
5	检查情况	20 min	计算机			
6	评价情况	20 min	课程伴侣			
班级		教师签字		日期		

表 10-12　成绩报告单

班级Input 输入系统学习场(课程)成绩报告单													
学习场 K	Input 输入系统												
学习情境 M	了解 Input 输入系统												
典型工作过程描述	创建项目—了解 Input 输入系统—根据 Input 输入系统创建一个完整项目								学时		4 学时(180 min)		
序号	姓名	第一个学习任务				第二个学习任务				第 N 个学习任务				总评
		自评 ×%	互评 ×%	教师评 ×%	合计	自评 ×%	互评 ×%	教师评 ×%	合计	自评 ×%	互评 ×%	教师评 ×%	合计	
1														
2														
3														
4														
5														
6														
7														
8														
9														
10														
11														
12														
13														
14														
15														
16														
17														
18														
19														
20														
21														
22														
23														
24														
25														
26														
27														
28														
班级		教师签字								日期				

10.2 理论指导

Input 输入系统 1

Input 输入系统 2

Input 为访问输入系统的接口。使用该类来读取传统游戏输入中设置的轴，以及访问移动设备上的多点触控/加速度计数据。要读取轴，请将 Input. GetAxis 与以下默认轴之一配合使用："Horizontal"和"Vertical"映射到游戏杆（A、D、W、S 和箭头键）。"Mouse X"和"Mouse Y"映射到鼠标增量。"Fire1""Fire2""Fire3"映射到 Cmd 键和三个鼠标或游戏杆按钮。可以添加新输入轴。若要使用输入来进行任何类型的移动行为，请使用 Input. GetAxis。它可提供平滑且可配置的输入，可以映射到键盘、游戏杆或鼠标。Input. GetButton 仅用于事件等操作，不要将它用于移动操作。Input. GetAxis 将使脚本代码更简洁。还有重要的一点，执行 Update 命令前，不会重置 Input 标志。建议在 Update 循环中进行所有的 Input 调用（图 10-1）。

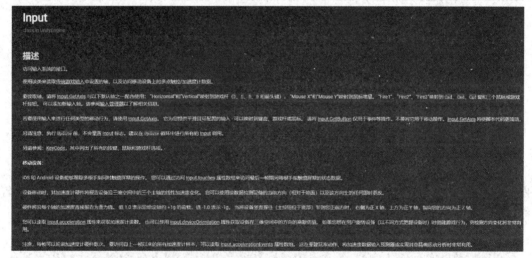

图 10-1

1. 虚拟按键

GetAxis 返回由 axisName 标识的虚拟轴的值，返回值为 float 类型（图 10-2）。

对于键盘和游戏杆输入，该值将处于 $-1 \sim 1$ 的范围内。如果轴设置为增量鼠标移动，则将鼠标增量乘以轴灵敏度，范围不为 $-1 \sim 1$。

该值与帧率无关；使用该值时，无须担心帧率变化问题。要设置输入或查看 axisName 的选项，请执行"Edit"→"Project Settings"→"Input"命令。这将调出 Input Manager。展开 Axis 可查看当前输入的列表。可以使用其中一个作为 /axisName/。要重命名输入或更改 Positive Button 等，展开其中一个选项，然后在 Name 字段或 Positive Button 字段中更改名称。此外，将 Type 更改为 Joystick Axis。要添加新的输入，将 Size 字段中的数字加 1（图 10-3）。

图 10-2

图 10-3

2. 按键脚本

(1)GetButton：当按住 buttonName 标识的虚拟按钮时，返回 true。

可以想象自动射击的场景，只要按住此按钮，该函数就一直返回 true。只在实现触发操作的事件(例如武器开火)时使用该函数。buttonName 参数通常为 InputManager 中的名称之一，例如 Jump 或 Fire1。松开按钮时，GetButton 将返回 false(图 10-4)。

图 10-4

(2)GetButtonDown：在用户按下由 buttonName 标识的虚拟按钮的帧期间返回 true (图 10-5)。

图 10-5

(3)GetButtonUp：在用户按下之后释放由 buttonName 标识的虚拟按钮的第一帧返回 true(图 10-6)。

图 10-6

（4）GetKey：在用户按下 name 标识的键时返回 true（图 10-7）。

图 10-7

（5）GetKeyDown：在用户开始按下 name 标识的键的帧期间返回 true（图 10-8）。

图 10-8

（6）GetKeyUp：在用户按下之后释放 name 标识的键的帧期间返回 true（图 10-9）。

图 10-9

（7）GetMouseButton：返回是否按下了给定的鼠标按钮。

button 值为 0 表示键盘上左箭头按钮，1 表示键盘上右箭头按钮，2 表示键盘上下箭头（中间）按钮。按下鼠标按钮时返回 true，释放时返回 false（图 10-10）。

图 10-10

（8）GetMouseButtonDown：在用户按下给定鼠标按钮的帧期间返回 true。

button 值为 0 表示键盘上左箭头按钮，1 表示键盘上右箭头按钮，2 表示键盘上下箭头（中间）按钮。按下鼠标按钮时返回 true，释放时返回 false（图 10-11）。

图 10-11

（9）GetMouseButtonUp：在按下之后用户释放给定鼠标按钮的帧期间返回 true。

button 值为 0 表示键盘上左箭头按钮，1 表示键盘上右箭头按钮，2 表示键盘上下箭头（中间）按钮。按下鼠标按钮时返回 true，释放时返回 false（图 10-12）。

Input.GetMouseButtonUp

图 10-12

项目11　UI系统

11.1　项目表单

表 11-1　学习性工作任务单

学习场 K	UI 系统				
学习情境 L	了解 UI 系统				
学习任务 M	创建一个简单的 UI 界面		学时		4 学时(180 min)
典型工作过程描述	创建项目—了解 UI 系统—创建一个简单的 UI 界面				
学习目标	1. 了解 UI 系统 2. 熟悉 UI 系统 3. 创建一个简单的 UI 界面				
任务描述	熟悉 UI 系统并创建一个简单的 UI 界面目				
学时安排	资讯 20 min	计划 10 min	决策 10 min	实施 100 min	检查 20 min	评价 20 min
对学生的要求	1. 安装好软件 2. 课前做好预习 3. 熟悉 UI 系统 4. 创建一个简单的 UI 界面				
参考资料	1. 素材包 2. 微视频 3. PPT				

表 11-2 资讯单

学习场 K	UI 系统		
学习情境 L	了解 UI 系统		
学习任务 M	创建一个简单的 UI 界面	学时	20 min
典型工作过程描述	创建项目—了解 UI 系统—创建一个简单的 UI 界面		
搜集资讯的方式	1. 教师讲解 2. 互联网查询 3. 同学交流		
资讯描述	查看教师提供的资料，获取信息，便于创建		
对学生的要求	1. 软件安装完成 2. 课前做好预习 3. 熟悉 UI 系统 4. 创建一个简单的 UI 界面		
参考资料	1. 素材包 2. 微视频 3. PPT		

表 11-3 计划单

学习场 K	UI 系统			
学习情境 L	了解 UI 系统			
学习任务 M	创建一个简单的 UI 界面		学时	10 min
典型工作过程描述	创建项目—了解 UI 系统—创建一个简单的 UI 界面			
计划制订的方式	同学间分组讨论			
序号	工作步骤		注意事项	
1	创建一个新的项目			
2	熟悉 UI 系统			
3	创建一个简单的 UI 界面			
计划评价	班级		第____组	组长签字	
	教师签字		日期		
	评语：				

表 11-4　决策单

学习场 K	UI 系统		
学习情境 L	了解 UI 系统		
学习任务 M	创建一个简单的 UI 界面	学时	10 min
典型工作过程描述	创建项目—了解 UI 系统—创建一个简单的 UI 界面		

			计划对比		
序号	计划的可行性	计划的经济性	计划的可操作性	计划的实施难度	综合评价
1					
2					
3					
4					
5					
6					
7					
8					
9					
10					

决策评价	班级		第＿＿＿组	组长签字	
	教师签字		日期		
	评语：				

表 11-5　实施单

学习场 K	UI 系统		
学习情境 L	了解 UI 系统		
学习任务 M	创建一个简单的 UI 界面	学时	100 min
典型工作过程描述	创建项目—了解 UI 系统—创建一个简单的 UI 界面		

序号	实施步骤	注意事项
1	创建一个新的项目	新建项目
2	了解 UI 系统	UI 系统
3	创建一个简单的 UI 界面	是否合理，是否正确

实施说明：

1. 启动 UnityHub 程序后，需要创建一个项目
2. 熟悉 UI 系统
3. 创建一个简单的 UI 界面

实施评价	班级		第＿＿＿组	组长签字	
	教师签字		日期		
	评语：				

表 11-6　检查单

学习场 K	UI 系统			
学习情境 L	了解 UI 系统			
学习任务 M	创建一个简单的 UI 界面		学时	20 min
典型工作过程描述	创建项目—了解 UI 系统—创建一个简单的 UI 界面			
序号	检查项目	检查标准	学生自查	教师检查
1	资讯环节	获取相关信息情况		
2	计划环节	熟悉 UI 系统		
3	实施环节	创建一个简单的 UI 界面		
4	检查环节	各个环节逐一检查		
检查评价	班级		第＿＿＿组	组长签字	
	教师签字		日期		
	评语：				

表 11-7　评价单

学习场 K	UI 系统			
学习情境 L	了解 UI 系统			
学习任务 M	创建一个简单的 UI 界面		学时	20 min
典型工作过程描述	创建项目—了解 UI 系统—创建一个简单的 UI 界面			
评价项目	评价子项目	学生自评	组内评价	教师评价
资讯环节	1. 听取教师讲解 2. 互联网查询情况 3. 同学交流情况			
计划环节	1. 查询资料情况 2. 熟悉 UI 系统			
实施环节	1. 学习态度 2. 对 UI 系统的掌握程度			
最终结果	综合情况			
评价	班级		第＿＿＿组	组长签字	
	教师签字		日期		
	评语：				

表 11-8　教学引导文设计单

学习场 K	UI 系统	学习情境 L	了解 UI 系统	参照系	信息工程学院
		学习任务 M	创建一个简单的 UI 界面		
普适性 工作过程 典型工 作过程	资讯	计划	决策	实施	检查	评价
分析 UI 系统	教师讲解	同学分组讨论	计划的可行性	了解 UI 系统	获取相关 信息情况	评价学习 态度
熟悉 UI 系统	自行掌握	熟悉 UI 系统	计划的 经济性	设置操作方式	检查组件	评价学生的 熟悉度
进行运用	创建一个简单 的 UI 界面	设计操作	计划的 实施难度	运用所学 所有知识	检查语句 运用是否正确	软件 熟练程度
保存项目文件	了解项目 文件的格式	了解项目 文件的格式	综合评价	保存项目文件	检查项目 文件的格式	评价项目

表 11-9　教学反馈单(学生反馈)

学习场 K	UI 系统			
学习情境 L	了解 UI 系统			
学习任务 M	创建一个简单的 UI 界面		学时	4 学时(180 min)
典型工作过程描述	创建项目—了解 UI 系统—创建一个简单的 UI 界面			
调查项目	序号	调查内容		理由描述
	1	资讯环节		
	2	计划环节		
	3	实施环节		
	4	检查环节		

您对本次课程教学的改进意见：

调查信息	被调查人姓名		调查日期	

表 11-10　分组单

学习场 K	UI 系统				
学习情境 L	了解 UI 系统				
学习任务 M	创建一个简单的 UI 界面		学时	4 学时(180 min)	
典型工作过程描述	创建项目—了解 UI 系统—创建一个简单的 UI 界面				
分组情况	组别	组长	组员		
	1				
	2				
	3				
	4				
	5				
	6				
	7				
	8				
分组说明					
班级		教师签字		日期	

表 11-11　教师实施计划单

学习场 K	UI 系统				
学习情境 L	了解 UI 系统				
学习任务 M	创建一个简单的 UI 界面		学时	4 学时(180 min)	
典型工作过程描述	创建项目—了解 UI 系统—创建一个简单的 UI 界面				
序号	工作与学习步骤	学时	使用工具	地点	方式	备注
1	资讯情况	20 min	互联网			
2	计划情况	10 min	计算机			
3	决策情况	10 min	计算机			
4	实施情况	100 min	Unity			
5	检查情况	20 min	计算机			
6	评价情况	20 min	课程伴侣			
班级		教师签字		日期		

表 11-12 成绩报告单

<table>
<tr><td colspan="15" align="center">_____班级 UI 系统学习场(课程)成绩报告单</td></tr>
<tr><td colspan="2">学习场 K</td><td colspan="13">UI 系统</td></tr>
<tr><td colspan="2">学习情境 M</td><td colspan="13">了解 UI 系统</td></tr>
<tr><td colspan="2">典型工作过程描述</td><td colspan="9">创建项目—了解 UI 系统—创建一个简单的 UI 界面</td><td>学时</td><td colspan="2">4 学时(180 min)</td></tr>
<tr><td rowspan="2">序号</td><td rowspan="2">姓名</td><td colspan="4" align="center">第一个学习任务</td><td colspan="4" align="center">第二个学习任务</td><td colspan="4" align="center">第 N 个学习任务</td><td rowspan="2">总评</td></tr>
<tr><td>自评 ×%</td><td>互评 ×%</td><td>教师评 ×%</td><td>合计</td><td>自评 ×%</td><td>互评 ×%</td><td>教师评 ×%</td><td>合计</td><td>自评 ×%</td><td>互评 ×%</td><td>教师评 ×%</td><td>合计</td></tr>
<tr><td>1</td><td></td><td></td><td></td><td></td><td></td><td></td><td></td><td></td><td></td><td></td><td></td><td></td><td></td><td></td></tr>
<tr><td>2</td><td></td><td></td><td></td><td></td><td></td><td></td><td></td><td></td><td></td><td></td><td></td><td></td><td></td><td></td></tr>
<tr><td>3</td><td></td><td></td><td></td><td></td><td></td><td></td><td></td><td></td><td></td><td></td><td></td><td></td><td></td><td></td></tr>
<tr><td>4</td><td></td><td></td><td></td><td></td><td></td><td></td><td></td><td></td><td></td><td></td><td></td><td></td><td></td><td></td></tr>
<tr><td>5</td><td></td><td></td><td></td><td></td><td></td><td></td><td></td><td></td><td></td><td></td><td></td><td></td><td></td><td></td></tr>
<tr><td>6</td><td></td><td></td><td></td><td></td><td></td><td></td><td></td><td></td><td></td><td></td><td></td><td></td><td></td><td></td></tr>
<tr><td>7</td><td></td><td></td><td></td><td></td><td></td><td></td><td></td><td></td><td></td><td></td><td></td><td></td><td></td><td></td></tr>
<tr><td>8</td><td></td><td></td><td></td><td></td><td></td><td></td><td></td><td></td><td></td><td></td><td></td><td></td><td></td><td></td></tr>
<tr><td>9</td><td></td><td></td><td></td><td></td><td></td><td></td><td></td><td></td><td></td><td></td><td></td><td></td><td></td><td></td></tr>
<tr><td>10</td><td></td><td></td><td></td><td></td><td></td><td></td><td></td><td></td><td></td><td></td><td></td><td></td><td></td><td></td></tr>
<tr><td>11</td><td></td><td></td><td></td><td></td><td></td><td></td><td></td><td></td><td></td><td></td><td></td><td></td><td></td><td></td></tr>
<tr><td>12</td><td></td><td></td><td></td><td></td><td></td><td></td><td></td><td></td><td></td><td></td><td></td><td></td><td></td><td></td></tr>
<tr><td>13</td><td></td><td></td><td></td><td></td><td></td><td></td><td></td><td></td><td></td><td></td><td></td><td></td><td></td><td></td></tr>
<tr><td>14</td><td></td><td></td><td></td><td></td><td></td><td></td><td></td><td></td><td></td><td></td><td></td><td></td><td></td><td></td></tr>
<tr><td>15</td><td></td><td></td><td></td><td></td><td></td><td></td><td></td><td></td><td></td><td></td><td></td><td></td><td></td><td></td></tr>
<tr><td>16</td><td></td><td></td><td></td><td></td><td></td><td></td><td></td><td></td><td></td><td></td><td></td><td></td><td></td><td></td></tr>
<tr><td>17</td><td></td><td></td><td></td><td></td><td></td><td></td><td></td><td></td><td></td><td></td><td></td><td></td><td></td><td></td></tr>
<tr><td>18</td><td></td><td></td><td></td><td></td><td></td><td></td><td></td><td></td><td></td><td></td><td></td><td></td><td></td><td></td></tr>
<tr><td>19</td><td></td><td></td><td></td><td></td><td></td><td></td><td></td><td></td><td></td><td></td><td></td><td></td><td></td><td></td></tr>
<tr><td>20</td><td></td><td></td><td></td><td></td><td></td><td></td><td></td><td></td><td></td><td></td><td></td><td></td><td></td><td></td></tr>
<tr><td>21</td><td></td><td></td><td></td><td></td><td></td><td></td><td></td><td></td><td></td><td></td><td></td><td></td><td></td><td></td></tr>
<tr><td>22</td><td></td><td></td><td></td><td></td><td></td><td></td><td></td><td></td><td></td><td></td><td></td><td></td><td></td><td></td></tr>
<tr><td>23</td><td></td><td></td><td></td><td></td><td></td><td></td><td></td><td></td><td></td><td></td><td></td><td></td><td></td><td></td></tr>
<tr><td>24</td><td></td><td></td><td></td><td></td><td></td><td></td><td></td><td></td><td></td><td></td><td></td><td></td><td></td><td></td></tr>
<tr><td>25</td><td></td><td></td><td></td><td></td><td></td><td></td><td></td><td></td><td></td><td></td><td></td><td></td><td></td><td></td></tr>
<tr><td>26</td><td></td><td></td><td></td><td></td><td></td><td></td><td></td><td></td><td></td><td></td><td></td><td></td><td></td><td></td></tr>
<tr><td>27</td><td></td><td></td><td></td><td></td><td></td><td></td><td></td><td></td><td></td><td></td><td></td><td></td><td></td><td></td></tr>
<tr><td>28</td><td></td><td></td><td></td><td></td><td></td><td></td><td></td><td></td><td></td><td></td><td></td><td></td><td></td><td></td></tr>
<tr><td colspan="2">班级</td><td colspan="4">教师签字</td><td colspan="5"></td><td colspan="2">日期</td><td></td></tr>
</table>

11.2 理论指导

UI 系统可用于快速直观地创建用户界面。本部分将介绍 UI 系统的主要功能(图 11-1)。

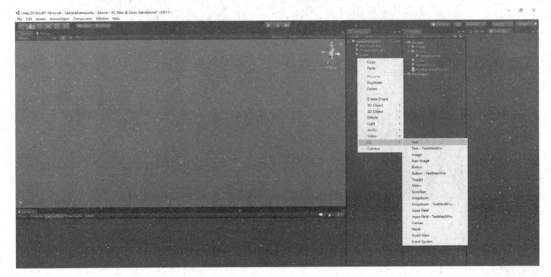

图 11-1

1. 画布(Canvas)概述

画布（Canvas）是应该容纳所有 UI 元素的区域。画布是一种带有画布组件的游戏对象，所有的 UI 元素都必须是 Canvas 的子项。创建新的 UI 元素（如执行菜单"GameObject"→"UI"→"Image"命令创建图像）时，如果场景中还没有画布，则会自动创建画布。UI 元素将创建为此画布的子项。不过一个场景中并不是只能存在一个画布，所以可以根据需求使用多个画布。

画布区域在 Scene 视图中显示为矩形。这样可以轻松定位 UI 元素，而无须始终显示 Game 视图。画布会使用 EventSystem 对象来协助消息系统。EventSystem 的作用就是用来处理场景中 UI 的交互事件(图 11-2)。

2. 渲染模式

画布具有渲染模式（Render Mode）设置，可用于屏幕空间或世界空间中进行渲染。

(1)Screen Space—Overlay(Canvas 的默认 Render Mode 为 Screen Space—Overlay)。此渲染模式将 UI 元素放置于在场景之上渲染的屏幕。如果调整屏幕大小或更改分辨率，则画布将自动更改大小来适应此情况。在该模式下将无法手动的调整 Canvas 尺寸等信息，因为 Canvas 的大小是由 Game 视图的大小决定，而 Canvas 是覆盖在整个 Game 视图之上的。

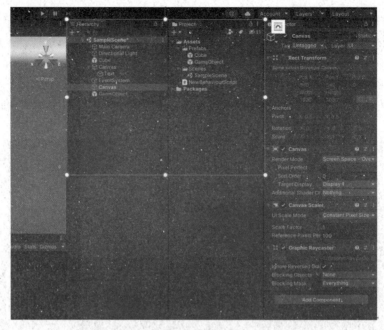

图 11-2

（2）Screen Space—Camera。此渲染模式类似 Screen Space—Overlay，但在此模式下，画布放置在指定摄像机前面的给定距离处。UI 元素由此摄像机渲染，这意味着摄像机设置会影响 UI 的外观。如果摄像机设置为正交视图，则 UI 元素将以透视图渲染，透视失真量可由摄像机视野控制。如果调整屏幕大小、更改分辨率或摄像机视锥体发生改变，则画布也将自动更改大小来适应此情况。

（3）World Space。在此渲染模式下，画布的行为与场景中的所有其他对象相同。画布大小可用矩形变换进行手动设置，而 UI 元素将基于 3D 位置在场景中的其他对象前面或后面渲染。此模式对于要成为世界一部分的 UI 非常有用。这种界面也称为"叙事界面"。需要注意的是，在 VR 项目的开发中，所有的 Canvas 必须设置为 World Space。

3. UI—text（文本）

文本组件也称为标签（Label），有一个文本区域可用于输入要显示的文本，可以设置字体、字体样式、字体大小以及文本是否支持副文本功能。有一些选项可以设置文本的对齐方式、水平和垂直溢出的设置（控制文本大于矩形的宽度或高度时会发生什么情况）以及一个使文本调整大小来适应可用空间的 Best Fit 选项。

UI 系统之 Text

创建 text 在 Hierarchy 面板中单击右键执行"UI"→"text"命令（图 11-3）。

主要有以下几个属性：

（1）Text。控件显示的文本。

（2）Character 字符。

①Font：用于显示文本的字体。

②Font Style：应用于文本的样式。选项包括 Normal（正常）、Bold（加粗）、Italic（斜体）和 Bold and Italic（加粗斜体）。

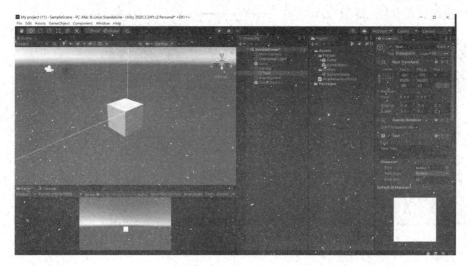

图 11-3

③Font Size：显示的文本的大小。

④Line Spacing：文本行之间的垂直间距。

⑤Rich Text：文本中的标记元素是否应解释为副文本样式，可以显示 HTML 标签的效果。如果对 HTML 标签感兴趣，可自行网上查询。

（3）Paragraph 段落。

①Alignment：文本的水平和垂直对齐方式。

②Align by Geometry：使用字形几何形状的范围（而不是字形指标）执行水平对齐。

③Horizontal Overflow：用于处理文本太宽而无法放入矩形内的情况的方法。提供的选项为 Wrap 和 Overflow。

④Vertical Overflow：用于处理换行文本太高而无法放入矩形内的情况的方法。提供的选项为 Truncate 和 Overflow。

⑤Best FitUnity：应该忽略大小属性并尝试直接将文本放入控件的矩形。

⑥Color：用于渲染文本的颜色。

⑦Material：用于渲染文本的材质（图 11-4）。

4. UI—Button

按钮有一个 OnClick UnityEvent 用于定义单击按钮时将执行的操作。Button 元素默认会有一个 Text 元素作为子对象（图 11-5）。

图 11-4

(1)组件属性介绍(图 11-6)。

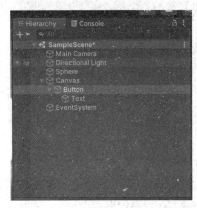
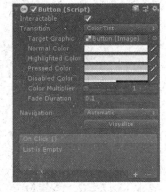

UI 系统之 Button

图 11-5 图 11-6

①Interactable：勾选，按钮可用；取消勾选，按钮不可用。

②Transition：按钮在状态改变时自身的过渡方式。默认为 Color Tint(颜色改变)、Sprite Swap(图片切换)、Animation(执行动画)

③Normal Color(默认颜色)：初始状态的颜色。

④Highlighted Color(高亮颜色)：选中状态或是鼠标指针靠近会进入高亮状态。

⑤Pressed Color(按下颜色)：鼠标单击或是按钮处于选中状态时按下 Enter 键。

⑥Disabled Color(禁用颜色)：禁用时颜色。

⑦Color Multiplier(颜色切换系数)：颜色切换速度，越大，则颜色在几种状态间变化速度越快。

⑧Fade Duration(衰落延时)：颜色变化的延时时间，越大，则变化越不明显。

当选择 Sprite Swap，出现的信息可这样设置：

①Highlighted Sprite(高亮图片)：选中状态或是鼠标指针靠近会进入高亮状态。

②Pressed Sprite(按下图片)：鼠标单击或是按钮处于选中状态时按下 Entcr 键。

③Disabled Sprite(禁用图片)：禁用时图片。

(2)按键触发的步骤如下(图 11-7)：

图 11-7

①创建一个公共的方法(图 11-8)。

②将脚本挂在场景中游戏物体身上(图 11-9)。

③on Click 选项中按＋键(图 11-10)。

④将带有脚本的物体拖入对应位置(图 11-11)。

⑤在 Function 中选择刚刚创建的方法(图 11-12)。

这样就可以触发相应的方法(图 11-13)。

```
        // Start is called before the first frame update
        void Start()
        {

        }

        // Update is called once per frame
        void Update()
        {

        }

        public void ButtonAction()
        {

        }
```

图 11-8

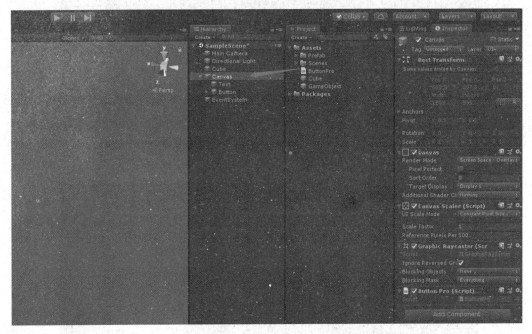

图 11-9

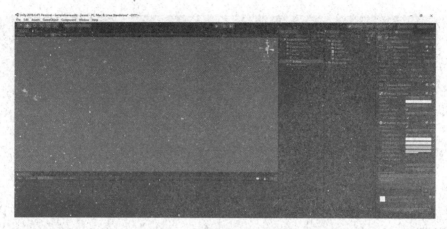

图 11-10

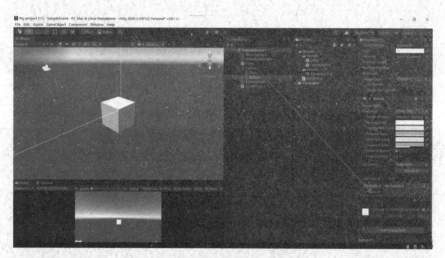

图 11-11

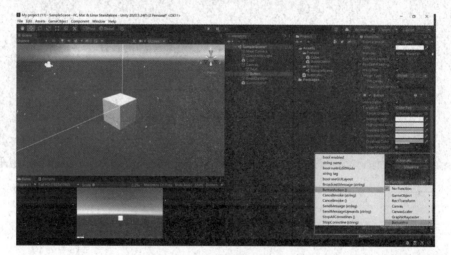

图 11-12

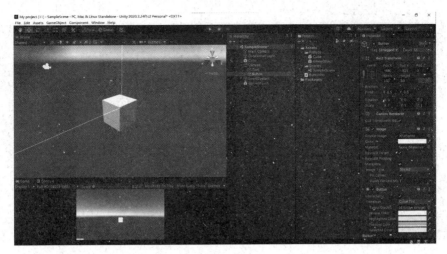

图 11-13

5. UI—Image

图像控件向用户显示非交互式图像。此图像可用于装饰、图标等，也可以从脚本更改图像以便反映其他控件的更改。该控件类似原始图像（Raw Image)控件，但为动画化图像和准确填充控件矩形提供了更多选项。图像控件要求其纹理为精灵，而原始图像可以接受任何纹理。

UI 系统之 Image

组件属性介绍(图 11-14)。

(1)Source Image：表示要显示的图像的纹理(必须作为精灵导入)。

(2)Color：应用于图像的颜色。

(3)Material：用于渲染图像的材质。

(4) Raycast Target：是否应将此图像视为射线投射目标。

图 11-14

(5)Preserve Aspect：确保图像保持现有尺寸。

(6)Set Native Size：使用此按钮可将图像框的尺寸设置为纹理的原始像素大小。

6. UI—Panel

Panel 控件又称为面板，面板实际上就是一个容器，在其上可放置其他 UI 控件。当移动面板时，放在其中的 UI 控件就会跟随移动，这样可以更加合理与方便地移动与处理一组控件。拖动面板控件的 4 个角或 4 条边可以调节面板的大小(图 11-15)。

7. Toggle

Toggle 元素的作用就是一个开关。只有当 Toggle 开启时，才能单击"Button"按钮，否则不允许单击"Button"按钮。在 Canvas 中添加 Toggle 元素，如图 11-16 所示。

在 Canvas 下添加 Toggle 元素。Toggle 元素有两个子对象，分别是 Background 和 Label，两者用于控制 Toggle 的外观(图 11-17)。

图 11-15

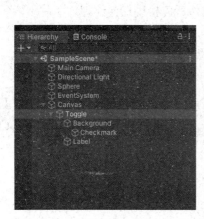

图 11-16

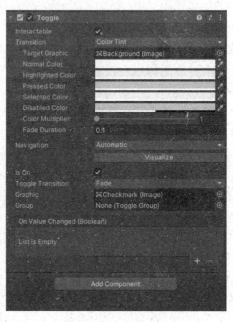

图 11-17

8. Slider

Slider 滑动条，可用于控制音量等。在 Canvas 中添加 Slider 元素，如图 11-18 所示。

Slider 元素是由多个 image 元素组成的，可通过更换 Image 中的图片来更改 Slider 样式(图 11-19)。

9. LayoutGroup 组件

LayoutGroup 组件的功能是控制子物体的尺寸和位置。LayoutGroup 组件包括 Horizontal Layout Group(横向排列子物体)、Vertical Layout Group(纵向排列子物体)和 Grid Layout Group(网格排列子物体)三个组件。LayoutGroup 组件并不控制自身的布局属性，可以使用上面提到的组件自动控制或手动设置。LayoutGroup 组件可以任意组合嵌套。

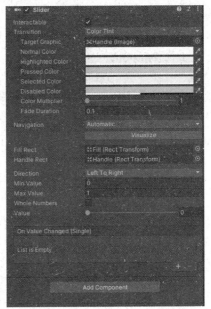

图 11-18　　　　　　　　　　　　　　　图 11-19

（1）Horizontal Layout Group（横向排列子物体）（图 11-20）。

①Padding：内边距。

②Spacing：元素之间的间距。

③Child Alignment：子物体的对齐方式。

④Child Controls Size：子物体是否自己控制自己的尺寸。

⑤Child Force Expand：是否强制子物体拉伸来充满所有的可用空间。

（2）Vertical Layout Group（纵向排列子物体）规则和 Horizontal Layout Group 相似（图 11-21）。

图 11-20　　　　　　　　　　　　　图 11-21

（3）Grid Layout Group（网格排列子物体）。Grid Layout Group 将子物体网格布局。

与横向/纵向排列不同，Grid 排列忽略子物体的 minimum、preferred、flexible 属性，使用 Cell Size 来设置它们的尺寸（图 11-22）。

图 11-22

①Padding：内边距。

②Cell Size：子物体的尺寸。

③Spacing：子物体之间的间距。

④Start Corner：第一个元素布局的位置。

⑤Start Axis：布局的方向，是横向还是竖向。

⑥Child Alignment：子物体对其的方式。

⑦Constraint：给 Grid 设置一些约束(股东行数/列数)，来辅助自动布局系统。

项目12　射线Ray

12.1　项目表单

表 12-1　学习性工作任务单

学习场 K	射线 Ray					
学习情境 L	了解射线 Ray					
学习任务 M	创建射线 Ray 并加入关于射线脚本			学时		4 学时(180 min)
典型工作过程描述	创建项目—了解射线 Ray—创建射线 Ray 并加入关于射线脚本					
学习目标	1. 了解射线 Ray 2. 熟悉射线 Ray 3. 创建射线 Ray 并加入关于射线脚本					
任务描述	熟悉射线 Ray 且创建射线 Ray 并加入关于射线脚本					
学时安排	资讯 20 min	计划 10 min	决策 10 min	实施 100 min	检查 20 min	评价 20 min
对学生的要求	1. 安装好软件 2. 课前做好预习 3. 熟悉射线 Ray 4. 创建射线 Ray 并加入关于射线脚本					
参考资料	1. 素材包 2. 微视频 3. PPT					

表 12-2　资讯单

学习场 K	射线 Ray		
学习情境 L	了解射线 Ray		
学习任务 M	创建射线 Ray 并加入关于射线脚本	学时	20 min
典型工作过程描述	创建项目—了解射线 Ray—创建射线 Ray 并加入关于射线脚本		
搜集资讯的方式	1. 教师讲解 2. 互联网查询 3. 同学交流		
资讯描述	查看教师提供的资料，获取信息		
对学生的要求	1. 软件安装完成 2. 课前做好预习 3. 熟悉射线 Ray 4. 创建射线 Ray 并加入关于射线脚本		
参考资料	1. 素材包 2. 微视频 3. PPT		

表 12-3　计划单

学习场 K	射线 Ray		
学习情境 L	了解射线 Ray		
学习任务 M	创建射线 Ray 并加入关于射线脚本	学时	10 min
典型工作过程描述	创建项目—了解射线 Ray—创建射线 Ray 并加入关于射线脚本		
计划制订的方式	同学间分组讨论		
序号	工作步骤		注意事项
1	创建一个新的项目		
2	熟悉射线 Ray		
3	创建射线 Ray 并加入关于射线脚本		
计划评价	班级		第____组	组长签字	
	教师签字		日期		
	评语：				

表 12-4　决策单

学习场 K	射线 Ray		
学习情境 L	了解射线 Ray		
学习任务 M	创建射线 Ray 并加入关于射线脚本	学时	10 min
典型工作过程描述	创建项目—了解射线 Ray—创建射线 Ray 并加入关于射线脚本		

		计划对比			
序号	计划的可行性	计划的经济性	计划的可操作性	计划的实施难度	综合评价
1					
2					
3					
4					
5					
6					
7					
8					
9					
10					

	班级		第_____组	组长签字	
	教师签字		日期		
决策评价	评语：				

表 12-5　实施单

学习场 K	射线 Ray		
学习情境 L	了解射线 Ray		
学习任务 M	创建射线 Ray 并加入关于射线脚本	学时	100 min
典型工作过程描述	创建项目—了解射线 Ray—创建射线 Ray 并加入关于射线脚本		

序号	实施步骤	注意事项
1	创建一个新的项目	新建项目
2	了解射线 Ray	射线 Ray
3	创建射线 Ray 并加入关于射线脚本	是否合理，是否正确

实施说明：
1. 启动 UnityHub 程序后，需要创建一个项目
2. 熟悉射线 Ray
3. 创建射线 Ray 并加入关于射线脚本

	班级		第_____组	组长签字	
	教师签字		日期		
实施评价	评语：				

表 12-6　检查单

学习场 K	射线 Ray			
学习情境 L	了解射线 Ray			
学习任务 M	创建射线 Ray 并加入关于射线脚本		学时	20 min
典型工作过程描述	创建项目—了解射线 Ray—创建射线 Ray 并加入关于射线脚本			
序号	检查项目	检查标准	学生自查	教师检查
1	资讯环节	获取相关信息情况		
2	计划环节	熟悉射线 Ray		
3	实施环节	创建射线 Ray 并加入关于射线脚本		
4	检查环节	各个环节逐一检查		
检查评价	班级		第____组	组长签字	
	教师签字		日期		
	评语:				

表 12-7　评价单

学习场 K	射线 Ray			
学习情境 L	了解射线 Ray			
学习任务 M	创建射线 Ray 并加入关于射线脚本		学时	20 min
典型工作过程描述	创建项目—了解射线 Ray—创建射线 Ray 并加入关于射线脚本			
评价项目	评价子项目	学生自评	组内评价	教师评价
资讯环节	1. 听取教师讲解 2. 互联网查询情况 3. 同学交流情况			
计划环节	1. 查询资料情况 2. 熟悉射线 Ray			
实施环节	1. 学习态度 2. 对射线 Ray 的掌握程度			
最终结果	综合情况			
评价	班级		第____组	组长签字	
	教师签字		日期		
	评语:				

表 12-8　教学引导文设计单

学习场 K	射线 Ray	学习情境 L	了解射线 Ray	参照系	信息工程学院
		学习任务 M	创建射线 Ray 并加入关于射线脚本		

典型工作过程 ＼ 普适性工作过程	资讯	计划	决策	实施	检查	评价
分析射线 Ray	教师讲解	同学分组讨论	计划的可行性	了解射线 Ray	获取相关信息情况	评价学习态度
熟悉射线 Ray	自行掌握	熟悉射线 Ray	计划的经济性	设置操作方式	检查组件	评价学生的熟悉度
进行运用	创建射线 Ray 并加入关于射线脚本	设计操作	计划的实施难度	运用所学所有知识	检查语句运用是否正确	软件熟练程度
保存项目文件	了解项目文件的格式	了解项目文件的格式	综合评价	保存项目文件	检查项目文件的格式	评价项目

表 12-9　教学反馈单(学生反馈)

学习场 K	射线 Ray		
学习情境 L	了解射线 Ray 系统		
学习任务 M	创建射线 Ray 并加入关于射线脚本	学时	4 学时(180 min)
典型工作过程描述	创建项目—了解射线 Ray—创建射线 Ray 并加入关于射线脚本		
调查项目	序号	调查内容	理由描述
	1	资讯环节	
	2	计划环节	
	3	实施环节	
	4	检查环节	

您对本次课程教学的改进意见：

调查信息	被调查人姓名		调查日期	

表 12-10　分组单

学习场 K	射线 Ray			
学习情境 L	了解射线 Ray			
学习任务 M	创建射线 Ray 并加入关于射线脚本		学时	4 学时(180 min)
典型工作过程描述	创建项目—了解射线 Ray—创建射线 Ray 并加入关于射线脚本			
分组情况	组别	组长	组员	
	1			
	2			
	3			
	4			
	5			
	6			
	7			
	8			
分组说明				
班级		教师签字		日期

表 12-11　教师实施计划单

学习场 K	射线 Ray				
学习情境 L	了解射线 Ray				
学习任务 M	创建射线 Ray 并加入关于射线脚本		学时	4 学时(180 min)	
典型工作过程描述	创建项目—了解射线 Ray—创建射线 Ray 并加入关于射线脚本				
序号	工作与学习步骤	学时	使用工具	地点	方式	备注
1	资讯情况	20 min	互联网			
2	计划情况	10 min	计算机			
3	决策情况	10 min	计算机			
4	实施情况	100 min	Unity			
5	检查情况	20 min	计算机			
6	评价情况	20 min	课程伴侣			
班级		教师签字		日期	

表 12-12　成绩报告单

班级射线 Ray 学习场(课程)成绩报告单															
学习场 K	射线 Ray														
学习情境 M	了解射线 Ray														
典型工作过程描述	创建项目—了解射线 Ray—创建射线 Ray 并加入关于射线脚本								学时		4 学时(180 min)				
序号	姓名	第一个学习任务				第二个学习任务				第 N 个学习任务				总评		
		自评 ×%	互评 ×%	教师评 ×%	合计	自评 ×%	互评 ×%	教师评 ×%	合计	自评 ×%	互评 ×%	教师评 ×%	合计			
1																
2																
3																
4																
5																
6																
7																
8																
9																
10																
11																
12																
13																
14																
15																
16																
17																
18																
19																
20																
21																
22																
23																
24																
25																
26																
27																
28																
班级		教师签字							日期							

12. 2　理论指导

射线表示形式，射线是从 origin 开始并按照某个 direction 行进的无限长的线(图 12-1)。

(1)射线的使用。Ray 射线类和 RaycastHit 射线投射碰撞信息类是两个最常用的射线工具类。

RaycastHit 用于从射线投射获取信息的结构。

其属性如下：

图 12-1

①barycentricCoordinate：命中的三角形的重心坐标。

②collider：命中的 Collider。

③distance：从射线原点到撞击点的距离。

④lightmapCoord：撞击点处的 UV 光照贴图坐标。

⑤normal：射线命中的表面的法线。

⑥point：世界空间中射线命中碰撞体的撞击点。

⑦rigidbody：命中的碰撞体的 Rigidbody。如果该碰撞体未附加到刚体，则值为 null。

⑧textureCoord：碰撞位置处的 UV 纹理坐标。

⑨textureCoord2：撞击点处的辅助 UV 纹理坐标。

⑩transform：命中的刚体或碰撞体的 Transform。

⑪triangleIndex：命中的三角形的索引。

射线 **Ray** 之鼠标
点击控制角色移动

(2)发射方法(Physics. Raycast)。向场景中的所有碰撞体投射一条射线，该射线起点为 origin，朝向 direction，长度为 maxDistance。可以选择提供一个 LayerMask，以过滤掉不想生成与其碰撞的碰撞体(图 12-2)。

图 12-2

(3)详细代码如下(图 12-3)：

```
using System. Collections;
using System. Collections. Generic;
using UnityEngine;

public class RayPro  : MonoBehaviour
{
```

射线 **Ray** 之鼠标
点击选择物体

```
Ray ray;
RaycastHit hit;
void Start()
{
ray= new Ray(this. transform. position. Vector3. forward);
  }
    // Update is called once per frame
  void Update()
  {
    If(Physics. Raycast(ray. out hit))
    {
      Debug. Log(hit. transform. name);
}
    }
}
```

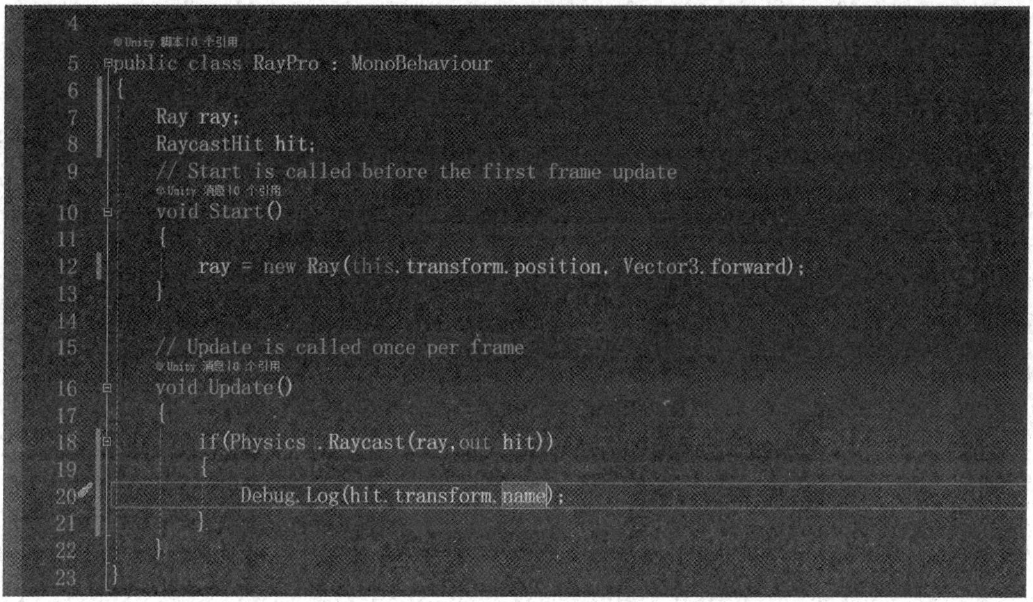

图 12-3

项目13 寻路系统

13.1 项目表单

表 13-1 学习性工作任务单

学习场 K	寻路系统		
学习情境 L	了解寻路系统		
学习任务 M	创建一个物体的寻路系统	学时	4 学时(180 min)
典型工作过程描述	创建项目—了解寻路系统—创建一个物体的寻路系统		
学习目标	1. 了解寻路系统 2. 熟悉寻路系统 3. 创建一个物体的寻路系统		
任务描述	熟悉寻路系统并创建一个物体的寻路系统		
学时安排	资讯 20 min　计划 10 min　决策 10 min　实施 100 min　检查 20 min　评价 20 min		
对学生的要求	1. 安装好软件 2. 课前做好预习 3. 熟悉寻路系统 4. 创建一个物体的寻路系统		
参考资料	1. 素材包 2. 微视频 3. PPT		

表 13-2　资讯单

学习场 K	寻路系统		
学习情境 L	了解寻路系统		
学习任务 M	创建一个物体的寻路系统	学时	20 min
典型工作过程描述	创建项目—了解寻路系统—创建一个物体的寻路系统		
搜集资讯的方式	1. 教师讲解 2. 互联网查询 3. 同学交流		
资讯描述	查看教师提供的资料，获取信息，便于创建		
对学生的要求	1. 软件安装完成 2. 课前做好预习 3. 熟悉寻路系统 4. 创建一个物体的寻路系统		
参考资料	1. 素材包 2. 微视频 3. PPT		

表 13-3　计划单

学习场 K	寻路系统		
学习情境 L	了解寻路系统		
学习任务 M	创建一个物体的寻路系统	学时	10 min
典型工作过程描述	创建项目—了解寻路系统—创建一个物体的寻路系统		
计划制订的方式	同学间分组讨论		
序号	工作步骤	注意事项	
1	创建一个新的项目		
2	寻路系统		
3	创建一个物体的寻路系统		
计划评价	班级 / 第　　组 / 组长签字		
	教师签字 / 日期		
	评语：		

表 13-4　决策单

学习场 K	寻路系统		
学习情境 L	了解寻路系统		
学习任务 M	创建一个物体的寻路系统	学时	10 min
典型工作过程描述	创建项目—了解寻路系统—创建一个物体的寻路系统		

计划对比					
序号	计划的可行性	计划的经济性	计划的可操作性	计划的实施难度	综合评价
1					
2					
3					
4					
5					
6					
7					
8					
9					
10					

决策评价	班级		第＿＿＿组	组长签字	
	教师签字		日期		
	评语：				

表 13-5　实施单

学习场 K	寻路系统		
学习情境 L	了解寻路系统		
学习任务 M	创建一个物体的寻路系统	学时	100 min
典型工作过程描述	创建项目—了解寻路系统—创建一个物体的寻路系统		

序号	实施步骤	注意事项
1	创建一个新的项目	新建项目
2	了解寻路系统	寻路系统
3	创建一个物体的寻路系统	是否合理，是否正确

实施说明：

1. 启动 UnityHub 程序后，需要创建一个项目
2. 熟悉寻路系统
3. 创建一个物体的寻路系统

实施评价	班级		第＿＿＿组	组长签字	
	教师签字		日期		
	评语：				

表13-6 检查单

学习场 K	寻路系统			
学习情境 L	了解寻路系统			
学习任务 M	创建一个物体的寻路系统		学时	20 min
典型工作过程描述	创建项目—了解寻路系统—创建一个物体的寻路系统			
序号	检查项目	检查标准	学生自查	教师检查
1	资讯环节	获取相关信息情况		
2	计划环节	熟悉寻路系统		
3	实施环节	创建一个物体的寻路系统		
4	检查环节	各个环节逐一检查		
检查评价	班级		第____组	组长签字	
	教师签字		日期		
	评语：				

表13-7 评价单

学习场 K	寻路系统			
学习情境 L	了解寻路系统			
学习任务 M	创建一个物体的寻路系统		学时	20 min
典型工作过程描述	创建项目—了解寻路系统—创建一个物体的寻路系统			
评价项目	评价子项目	学生自评	组内评价	教师评价
资讯环节	1. 听取教师讲解 2. 互联网查询情况 3. 同学交流情况			
计划环节	1. 查询资料情况 2. 熟悉寻路系统			
实施环节	1. 学习态度 2. 对寻路系统掌握程度			
最终结果	综合情况			
评价	班级		第____组	组长签字	
	教师签字		日期		
	评语：				

表 13-8　教学引导文设计单

学习场 K	寻路系统	学习情境 L	了解寻路系统	参照系		信息工程学院
		学习任务 M	创建一个物体的寻路系统			
普适性 工作过程 典型工 作过程	资讯	计划	决策	实施	检查	评价
分析寻路系统	教师讲解	同学分组讨论	计划的可行性	了解寻路系统	获取相关 信息情况	评价学习 态度
熟悉寻路系统	自行掌握	熟悉寻路系统	计划的经济性	设置操作方式	检查组件	评价学生 的熟悉度
进行运用	创建一个物体 的寻路系统	设计操作	计划的实施难度	运用所学所 有知识	检查语句 运用是否正确	软件熟练 程度
保存项目文件	了解项目 文件的格式	了解项目 文件的格式	综合评价	保存项目文件	检查项目 文件的格式	评价项目

表 13-9　教学反馈单(学生反馈)

学习场 K	寻路系统			
学习情境 L	了解寻路系统			
学习任务 M	创建一个物体的寻路系统		学时	4 学时(180 min)
典型工作过程描述	创建项目—了解寻路系统—创建一个物体的寻路系统			
调查项目	序号	调查内容		理由描述
	1	资讯环节		
	2	计划环节		
	3	实施环节		
	4	检查环节		
您对本次课程教学的改进意见:				
调查信息	被调查人姓名		调查日期	

表 13-10 分组单

学习场 K	寻路系统				
学习情境 L	了解寻路系统				
学习任务 M	创建一个物体的寻路系统		学时	4 学时(180 min)	
典型工作过程描述	创建项目—了解寻路系统—创建一个物体的寻路系统				
分组情况	组别	组长		组员	
	1				
	2				
	3				
	4				
	5				
	6				
	7				
	8				
分组说明					
班级		教师签字		日期	

表 13-11 教师实施计划单

学习场 K	寻路系统				
学习情境 L	了解寻路系统				
学习任务 M	创建一个物体的寻路系统		学时	4 学时(180 min)	
典型工作过程描述	创建项目—了解寻路系统—创建一个物体的寻路系统				
序号	工作与学习步骤	学时	使用工具	地点	方式	备注
1	资讯情况	20 min	互联网			
2	计划情况	10 min	计算机			
3	决策情况	10 min	计算机			
4	实施情况	100 min	Unity			
5	检查情况	20 min	计算机			
6	评价情况	20 min	课程伴侣			
班级		教师签字		日期		

表 13-12 成绩报告单

班级寻路系统学习场(课程)成绩报告单														
学习场 K	寻路系统													
学习情境 M	了解寻路系统													
典型工作过程描述	创建项目—了解寻路系统—创建一个物体的寻路系统								学时		4 学时(180 min)			
序号	姓名	第一个学习任务				第二个学习任务				第 N 个学习任务				总评
		自评×%	互评×%	教师评×%	合计	自评×%	互评×%	教师评×%	合计	自评×%	互评×%	教师评×%	合计	
1														
2														
3														
4														
5														
6														
7														
8														
9														
10														
11														
12														
13														
14														
15														
16														
17														
18														
19														
20														
21														
22														
23														
24														
25														
26														
27														
28														
班级		教师签字							日期					

13.2　理论指导

（1）创建导航网格的步骤如下：

①选择应影响导航的场景几何体，即可行走表面和障碍物。

②选中 Navigation Static 复选框以便在导航网格烘焙过程中包括所选对象（图 13-1）。

③调整 Bake 设置以匹配代理大小。其中：Agent Radius：定义代理中心与墙壁或窗台的接近程度；Agent Height：定义代理可以达到的空间有多低；Max Slope：定义代理走上坡道的陡峭程度；Step Height：定义代理可以踏上的障碍物的高度（图 13-2）。

Nav Mesh Agent
基本寻路系统

图 13-1　　　　　　　　图 13-2

④单击 Bake 以构建导航网格（图 13-3）。

（2）导航网格代理（Nav Mesh Agent）。使用 Nav Mesh Agent 组件可创建在朝目标移动时能够彼此避开的角色。代理（Agent）使用导航网格来推断游戏世界，并知道如何避开彼此以及其他移动障碍物。寻路和空间推断是使用导航网格代理的脚本 API 进行处理的（图 13-4）。

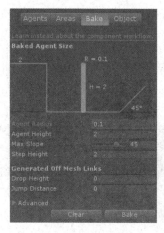

图 13-3　　　　　　　　图 13-4

①Agent Size：代理类型。

②Base offset：碰撞圆柱体相对于变换轴心点的偏移。

③Steering 包括：Speed：最大移动速度（以世界单位/秒表示）；Angular Speed：最大旋转速度（度/秒）；Acceleration：最大加速度（以世界单位/平方秒表示）；Stopping distance：当靠近目标位置的距离达到此值时，代理将停止；Auto Braking：启用此属性后，代理在到达目标时将减速。对于巡逻等行为（这种情况下，代理应在多个点之间平滑移动）应禁用此属性。

④Obstacle Avoidance 包括：Radius：代理的半径，用于计算障碍物与其他代理之间的碰撞。Height 代理通过头顶障碍物时所需的高度间隙。Quality：障碍躲避质量，如果拥有大量代理，则可以通过降低障碍躲避质量来节省 CPU 时间。如果将躲避设置为无，则只会解析碰撞，而不会尝试主动躲避其他代理和障碍物。Priority：执行避障时，此代理将忽略优先级较低的代理。该值应在 0～99 范围内，其中较低的数字表示较高的优先级。

⑤Path Finding 包括：Auto Traverse Of f-Mesh Link：设置为 true 可自动遍历网格外链接（Off-Mesh Link）。如果要使用动画或某种特定方式遍历网格外链接，则应关闭此功能。Auto Repath：启用此属性后，代理将在到达部分路径末尾时尝试再次寻路。当没有到达目标的路径时，将生成一条部分路径通向与目标最近的可达位置。Area Mask：描述了代理在寻路时将考虑的区域类型。在准备网格进行导航网格烘焙时，可设置每个网格区域类型。例如，可将楼梯标记为特殊区域类型，并禁止某些角色类型使用楼梯。

（3）寻路基本方法（NavMeshAgent）（图 13-5）。

①SetDestination 设置或更新目标，从而触发新路

NavMeshAgent
class in UnityEngine.AI / 继承自：Behaviour

图 13-5

径计算。请注意，路径可能在几帧之后才可用。计算路径时，pathPending 将为 true。如果有效路径可用，代理将恢复移动（图 13-6）。

图 13-6

②destination 获取代理在世界坐标系单位中的目标或尝试设置代理在其中的目标。如果设置了目标，但尚未处理路径，返回的位置将是与之前设置的位置最接近的有效导航网格位置；如果代理没有路径或请求路径，返回代理在导航网格上的位置；如果代理没有映射到导航网格（例如，场景中没有导航网格），返回无限远处的位置（图 13-7）。

（4）创建网格外链接。网格外链接（Off-Mesh Link）用于创建穿过可步行导航网格表面外部的路径。例如，跳过沟渠或围栏，或在通过门之前打开门，全都可

NavMeshAgent.destination
public Vector3 destination ;

图 13-7

以描述为网格外链接。我们将添加一个网格外链接组件来描述从上层平台到地面的跳跃（图 13-8）。

（5）导航网格障碍物（Nav Mesh Obstacle）。此组件允许描述导航网格代理在世界中导航时应避开的移动障碍物（例如，由物理系统控制的木桶或板条箱）。当障碍物正在移动时，导航网格代理会尽力避开它。当障碍物静止时，它会在导航网格中雕刻一个孔。导航网格代理随后将改变它们的路径以绕过障碍物，或者如果障碍物导致路径被完全阻挡，则寻找其他不同路线（图 13-9）。

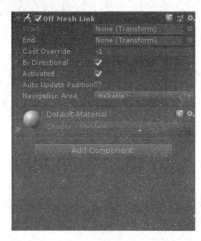

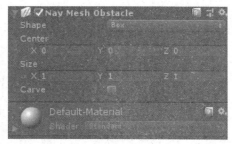

图 13-8　　　　　　　　　　　　　　　图 13-9

①Shape：障碍物几何体的形状，选择最适合对象形状的选项。包含两种形状，即 Box 和 Capsule。Box 需设置 Center 盒体的中心（相对于变换位置）和 Size 盒体的大小；Capsule 需设置 Center 胶囊体的中心（相对于变换位置）、Radius 胶囊体的半径和 Height 胶囊体的高度。

②Carve：勾选 Carve 复选框后，导航网格障碍物会在导航网格中创建一个孔。

③Move Threshold：当导航网格障碍物的移动距离超过 Move Threshold 设置的值时，Unity 会将其视为移动状态。使用此属性可设置该阈值距离来更新移动的雕孔。

④Time To Stationary：将障碍物视为静止状态所需等候的时间（以秒为单位）。

⑤Carve Only Stationary：启用此属性后，只有在静止状态时才会雕刻障碍物。

Nav Mesh Agent 断网格寻路系统

项目14 动画

14.1 项目表单

表 14-1 学习性工作任务单

学习场 K	动画					
学习情境 L	了解动画					
学习任务 M	创建动画			学时		4 学时(180 min)
典型工作过程描述	创建项目—了解动画—创建动画					
学习目标	1. 了解动画 2. 熟悉动画 3. 创建动画					
任务描述	熟悉动画并创建动画					
学时安排	资讯 20 min	计划 10 min	决策 10 min	实施 100 min	检查 20 min	评价 20 min
对学生的要求	1. 安装好软件 2. 课前做好预习 3. 熟悉动画 4. 创建动画					
参考资料	1. 素材包 2. 微视频 3. PPT					

表 14-2　资 讯 单

学习场 K	动画		
学习情境 L	了解动画		
学习任务 M	创建动画	学时	20 min
典型工作过程描述	创建项目—了解动画—创建动画		
搜集资讯的方式	1. 教师讲解 2. 互联网查询 3. 同学交流		
资讯描述	查看教师提供的资料，获取信息，便于创建		
对学生的要求	1. 软件安装完成 2. 课前做好预习 3. 熟悉动画 4. 创建动画		
参考资料	1. 素材包 2. 微视频 3. PPT		

表 14-3　计 划 单

学习场 K	动画			
学习情境 L	了解动画			
学习任务 M	创建动画		学时	10 min
典型工作过程描述	创建项目—了解动画—创建动画			
计划制订的方式	同学间分组讨论			
序号	工作步骤		注意事项	
1	创建一个新的项目			
2	熟悉动画			
3	创建动画			
计划评价	班级		第____组	组长签字
	教师签字		日期	
	评语：			

表 14-4   决策单

学习场 K	动画				
学习情境 L	了解动画				
学习任务 M	创建动画			学时	10 min
典型工作过程描述	创建项目—了解动画—创建动画				
计划对比					
序号	计划的可行性	计划的经济性	计划的可操作性	计划的实施难度	综合评价
1					
2					
3					
4					
5					
6					
7					
8					
9					
10					
决策评价	班级		第＿＿＿组		组长签字
	教师签字		日期		
	评语：				

表 14-5   实施单

学习场 K	动画		
学习情境 L	了解动画		
学习任务 M	创建动画	学时	100 min
典型工作过程描述	创建项目—了解动画—创建动画		
序号	实施步骤		注意事项
1	创建一个新的项目		新建项目
2	了解动画		动画
3	创建动画		是否合理，是否正确

实施说明：

1. 启动 UnityHub 程序后，需要创建一个项目
2. 熟悉动画
3. 创建动画

实施评价	班级		第＿＿＿组	组长签字
	教师签字		日期	
	评语：			

表 14-6 检查单

学习场 K	动画			
学习情境 L	了解动画			
学习任务 M	创建动画		学时	20 min
典型工作过程描述	创建项目—了解动画—创建动画			
序号	检查项目	检查标准	学生自查	教师检查
1	资讯环节	获取相关信息情况		
2	计划环节	熟悉动画		
3	实施环节	创建动画		
4	检查环节	各个环节逐一检查		
检查评价	班级		第____组	组长签字	
	教师签字		日期		
	评语:				

表 14-7 评价单

学习场 K	动画			
学习情境 L	了解动画			
学习任务 M	创建动画		学时	20 min
典型工作过程描述	创建项目—了解动画—创建动画			
评价项目	评价子项目	学生自评	组内评价	教师评价
资讯环节	1. 听取教师讲解 2. 互联网查询情况 3. 同学交流情况			
计划环节	1. 查询资料情况 2. 熟悉动画			
实施环节	1. 学习态度 2. 对动画的掌握程度			
最终结果	综合情况			
评价	班级		第____组	组长签字	
	教师签字		日期		
	评语:				

表 14-8　教学引导文设计单

学习场 K	动画	学习情境 L	了解动画	参照系	信息工程学院
		学习任务 M	创建动画		
普适性工作过程 典型工作过程	资讯	计划	决策	实施	检查	评价
分析动画	教师讲解	同学分组讨论	计划的可行性	了解动画	获取相关信息情况	评价学习态度
熟悉动画	自行掌握	熟悉动画	计划的经济性	设置操作方式	检查组件	评价学生的熟悉度
进行运用	创建动画	设计操作	计划的实施难度	运用所学所有知识	检查语句运用是否正确	软件熟练程度
保存项目文件	了解项目文件的格式	了解项目文件的格式	综合评价	保存项目文件	检查项目文件的格式	评价项目

表 14-9　教学反馈单(学生反馈)

学习场 K	动画		
学习情境 L	了解动画		
学习任务 M	创建动画	学时	4 学时(180 min)
典型工作过程描述	创建项目—了解动画—创建动画		
调查项目	序号	调查内容	理由描述
	1	资讯环节	
	2	计划环节	
	3	实施环节	
	4	检查环节	

您对本次课程教学的改进意见：

调查信息	被调查人姓名		调查日期	

表 14-10 分组单

学习场 K	动画			
学习情境 L	了解动画			
学习任务 M	创建动画		学时	4 学时(180 min)
典型工作过程描述	创建项目—了解动画—创建动画			
分组情况	组别	组长		组员
	1			
	2			
	3			
	4			
	5			
	6			
	7			
	8			
分组说明				
班级		教师签字		日期

表 14-11 教师实施计划单

学习场 K	动画					
学习情境 L	了解动画					
学习任务 M	创建动画		学时	4 学时(180 min)		
典型工作过程描述	创建项目—了解动画—创建动画					
序号	工作与学习步骤	学时	使用工具	地点	方式	备注
1	资讯情况	20 min	互联网			
2	计划情况	10 min	计算机			
3	决策情况	10 min	计算机			
4	实施情况	100 min	Unity			
5	检查情况	20 min	计算机			
6	评价情况	20 min	课程伴侣			
班级		教师签字		日期		

表 14-12　成绩报告单

序号	姓名	第一个学习任务				第二个学习任务				第 N 个学习任务				总评
		自评 ×%	互评 ×%	教师评 ×%	合计	自评 ×%	互评 ×%	教师评 ×%	合计	自评 ×%	互评 ×%	教师评 ×%	合计	
1														
2														
3														
4														
5														
6														
7														
8														
9														
10														
11														
12														
13														
14														
15														
16														
17														
18														
19														
20														
21														
22														
23														
24														
25														
26														
27														
28														
班级		教师签字							日期					

班级动画学习场（课程）成绩报告单

学习场 K	动画
学习情境 M	了解动画

典型工作过程描述	创建项目—了解动画—创建动画	学时	4 学时（180 min）

14.2 理论指导

Unity 的动画功能包括可重定向动画、运行时对动画权重的完全控制、动画播放中的事件调用、复杂的状态机层级视图和过渡、面部动画的混合形状等。

1. 动画系统概述

Unity 有一个丰富而复杂的动画系统(有时称为"Mecanim")。该系统具有以下功能:

(1)为 Unity 的所有元素(包括对象、角色和属性)提供简单工作流程和动画设置。

(2)支持导入的动画剪辑以及 Unity 内创建的动画。

(3)人形动画重定向能够将动画从一个角色模型应用到另一个角色模型。

(4)对齐动画剪辑的简化工作流程。

(5)方便预览动画剪辑以及它们之间的过渡和交互。因此,动画师与工程师之间的工作更加独立,使动画师能够在挂入游戏代码之前为动画构建原型并进行预览。

(6)提供可视化编程工具来管理动画之间的复杂交互。

(7)以不同逻辑对不同身体部位进行动画化。

(8)分层和遮罩功能。

2. 动画工作流程

Unity 的动画系统基于动画剪辑的概念形成。动画剪辑包含某些对象应如何随时间改变其位置、旋转或其他属性的相关信息。每个剪辑可视为单个线性录制,来自外部的动画剪辑由美术师或动画师使用第三方工具(例如 Autodesk © 3ds Max © 或 Autodesk © Maya ©)创建而成,或者来自动作捕捉工作室或其他来源。然后,动画剪辑将编入一个称为 Animator Controller 的类似流程图的结构化系统。Animator Controller 充当"状态机",负责跟踪当前应该播放哪个剪辑以及动画应该何时改变或混合在一起。

一个非常简单的 Animator Controller 可能只包含一个或两个剪辑,例如,使用此剪辑来控制能量块旋转和弹跳,或设置正确时间开门和关门的动画。一个更高级的 Animator Controller 可包含用于主角所有动作的几十段人形动画,并可能同时在多个剪辑之间进行混合,从而当玩家在场景中移动时提供流畅的动作。

Unity 的动画系统还具有用于处理人形角色的许多特殊功能。这些功能可让人形动画从任何来源(例如:动作捕捉、Asset Store 或某个其他第三方动画库)重定向到角色模型,并可调整肌肉定义。这些特殊功能由 Unity 的替身系统启用;在此系统中,人形角色会被映射到一种通用的内部格式。所有这些部分(动画剪辑、Animator Controller 和 Avatar)都通过 Animator 组件一起附加到某个游戏对象上。该组件引用了 Animator Controller,并(在必需时)引用此模型的 Avatar。Animator Controller 又进一步包含所使用的动画剪辑的引用。

3. 动画控制器

(1)Animator 组件。Animator 组件用于将动画分配给场景中的游戏对象。Animator 组件需要引用 Animator Controller,后者定义要使用哪

动画控制器

些动画剪辑，并控制何时以及如何在动画剪辑之间进行混合和过渡。如果游戏对象是具有 Avatar 定义的人形角色，还应在此组件中分配 Avatar，如图 14-1 所示。

图 14-1

①Controller：附加到此角色的 Animator Controller。

②Avatar：此角色的 Avatar(如果 Animator 用于对人形角色进行动画化)。

③Apply Root Motion：选择从动画本身还是从脚本控制角色的位置和旋转。

④Update Mode：此选项允许选择 Animator 何时更新以及应使用哪个时间标度。

⑤Normal Animator：与 Update 调用同步更新，Animator 的速度与当前时间标度匹配。如果时间标度变慢，动画将通过减速来匹配。

⑥Animate Physics Animator：与 FixedUpdate 调用同步更新(即与物理系统步调一致)。如果要对具有物理交互的对象(例如可四处推动刚体对象的角色)的运动进行动画化，应使用此模式。

⑦Unscaled Time Animator：与 Update 调用同步更新，但是 Animator 的速度忽略当前时间标度而不顾一切以 100% 速度进行动画化。此选项可用于以正常速度对 GUI 系统进行动画化，同时将修改的时间标度用于特效或暂停游戏。

⑧Culling Mode：可以为动画选择的剔除模式。

⑨Always Animate：始终进行动画化，即使在屏幕外也不要剔除。

⑩Cull Update Transforms：未显示渲染器时，禁用变换组件的重定向、IK(反向动力学)和写入。

⑪Cull Completely：未显示渲染器时，完全禁用动画。

(2)创建 Animator Controller。可从 Animator Controller 视图(执行"Window"→"Animation"→"Animator"命令)中查看并设置角色行为。可通过多种方式创建 Animator Controller：

①从 Project 视图中，执行"Create"→"Animator Controller"命令。

②在 Project 视图中右键单击并执行"Create"→"Animator Controller"命令。

③从 Assets 菜单中，执行"Assets"→"Create"→"Animator Controller"命令(图 14-2)。

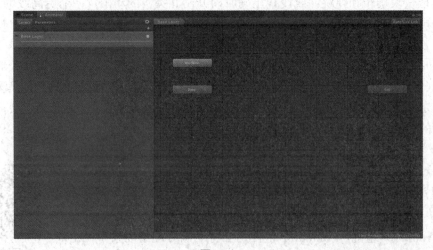

图 14-2

4. 动画过渡

状态机过渡有助于简化大型或复杂的状态机，允许对状态机逻辑进行更高级的抽象化。Animator 窗口中的每个视图都有一个进入（Entry）和退出（Exit）节点。在状态机过渡期间使用这些节点。过渡到状态机时使用进入节点。进入节点将接受评估，并根据设置的条件分支到目标状态。通

动画过渡

过此方式，进入节点可以通过在状态机启动时评估参数的状态来控制状态机的初始状态。

因为状态机始终具有默认状态，所以始终会有从进入节点分支到默认状态的默认过渡。

动画参数是在 Animator Controller 中定义的变量，可从脚本访问这些变量并向其赋值。这是脚本控制或影响状态机流程的方法。例如，可通过动画曲线更新参数的值，然后从脚本访问参数以便可改变音效的音高（就像它是一段动画一样）。同样，脚本可设置被 Mecanim 拾取的参数值。例如，脚本可设置参数来控制混合树。

可使用 Animator 窗口的 Parameters 部分来设置默认参数值（可在 Animator 窗口的右上角进行选择）。这些参数可分为 Float（带小数部分的数字）；Int（整数）；Bool（true 或 false 值，由复选框表示）；Trigger（当被过度使用时，由控制器重置的布尔值参数，以圆形按钮表示）四个基本类型，如图 14-3 所示。

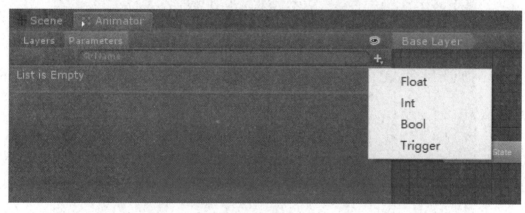

图 14-3

5. 程序控制

可使用以下 Animator 类中的函数，从脚本为参数赋值：SetFloat、SetInt、SetBool、SetTrigger 和 ResetTrigger。

其代码如下（图 14-4）：

程序控制

```
void Update()
    {
        float h= Input.GetAxis("Horizontal");
        float v= Input.GetAxis("Vertical");
        bool fire= Input.GetButtonDown("Fire1");
        animator.SetFloat("Strafe",h);
        ananimator.SetFloat("Forward",v);
```

```
        aimator.SetBool("Fire",fire);
}
void OnCollisionEnter(Collision col)
{
        if( col.gameObject.CompareTag("Enemy"))
        {
            animator.SetTrigger("Die");
        }
}
```

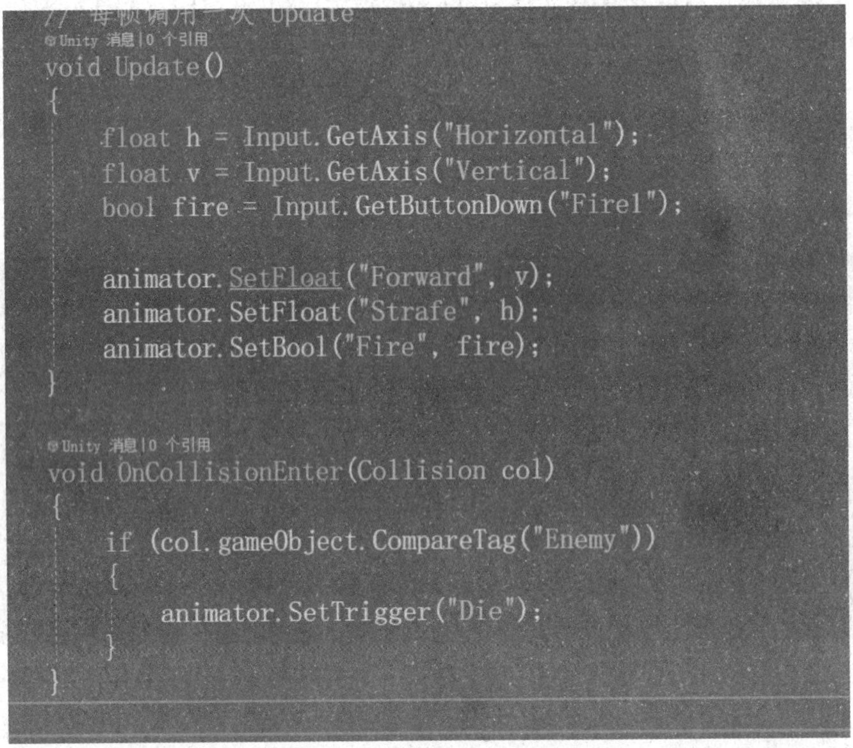

图 14-4

6. 动画层

Unity 使用动画层来管理不同身体部位的复杂状态机。相应的示例：有一个用于行走/跳跃的下身层，还有一个用于投掷物体/射击的上身层。可以从 Animator Controller 左上角的 Layers 小部件管理动画层(图 14-5)。单击窗口右侧的齿轮可显示该层的设置(图 14-6)。

在每一层上，可以指定遮罩(应用动画的动画模型的一部分)以及混合类型。Override 表示将忽略其他层的信息，而 Additive 表示将在先前层之上添加动画(可以通过按小部件上方的 + 来添加新层)。Mask 属性用于指定此层上使用的遮罩。例如，如果只想播放模型上半身的投掷动画，同时让角色也能够行走、奔跑或站立不动，则可以在层上使用一个遮罩，从而在定义上半身部分的位置播放投掷动画。

图 14-5

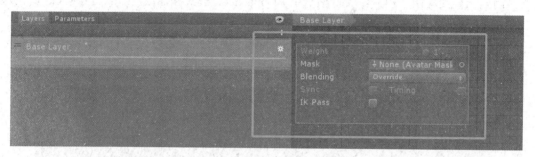

图 14-6

7. 动画制作(Animation)

Unity 中的 Animation 窗口可以直接在 Unity 内创建和修改动画剪辑，它旨在充当外部 3D 动画程序的强大而直接的替代方案。除了对运动进行动画化，编辑器还允许对材质和组件的变量进行动画化，并使用动画事件函数(在时间轴上的指定点调用这些函数)来充实动画剪辑。

(1)Animation 视图简介。Animation 视图用于预览和编辑 Unity 中已动画化的游戏对象的动画剪辑。要在 Unity 中打开 Animation 视图，请执行"Window"→"Animation"命令(图 14-7)。

图 14-7

（2）查看游戏对象上的动画。Animation 窗口与 Hierarchy 窗口、Project 窗口、Scene 视图和 Inspector 窗口相关联。与 Inspector 一样，Animation 窗口显示当前所选游戏对象或动画剪辑资源的动画时间轴和关键帧。可以使用 Hierarchy 窗口或 Scene 视图来选择游戏对象，或使用 Project 窗口来选择动画剪辑资源。

（3）属性。

①动画时间轴（图 14-8）。Animation 视图的右侧是当前剪辑的时间轴。每个动画属性的关键帧都显示在此时间轴

图 14-8

中。时间轴视图有两个模式：关键帧清单（Dopesheet）模式和曲线（Curves）模式。要在这些模式之间切换，请单击动画属性列表区域底部的 Dopesheet 或 Curve。

②时间轴（图 14-9）。关键帧清单模式提供更紧凑视图，允许在单个水平轨道中查看每个属性的关键帧序列。因此，可以查看多个属性或游戏对象的关键帧时间的概况。

图 14-9

③回放和帧导航的控制（图 14-10）。

从左到右，这些控制按钮主要包括预览模式（切换开/关）；录制模式（切换开/关），注意如果打开录制模式，则预览模式也会始终打开；

图 14-10

将回放头移到剪辑的开头；将回放头移到上一关键帧；播放动画；将回放头移到下一关键帧；将回放头移到剪辑的结尾。

还可使用以下键盘快捷键来控制回放：按逗号键跳到上一帧；按句号键跳到下一帧；按住 Alt 并按下逗号键跳到上一关键帧；按住 Alt 并按下句号键跳到下一关键帧。

8. Blend Tree 动画混合

游戏动画中的一项常见任务是为了角色之间动作更加自然流畅在两个或更多相似运动之间进行混合。最熟知的示例就是根据角色的速度来混合行走和奔跑动画；另一个示例是角色在奔跑期间转向时，向左或向右倾斜。

区分过渡（Transitions）与混合树（Blend Trees）十分重要。虽然两者都用于创建平滑动画，但它们用于不同种类的情况。

过渡用于在给定时间内从一个动画状态平滑过渡到另一动画状态。过渡作为动画状态机的一部分指定。如果过渡很快，从一个运动到完全不同运动的过渡通常没有问题。

混合树允许通过不同程度合并多个动画来使动画平滑混合。每个运动对最终效果的影响由一个混合参数控制，该参数只是与 Animator Controller 相关联的数字动画参数之一。为了使混合后的运动合理，要混合的运动必须具有相似的性质和时机。混合树是动画状态机中的一种特殊状态类型。

使用新混合树的步骤如下：

（1）右键单击 Animator Controller 窗口上的空白区域。

（2）从显示的上下文菜单中，执行"Create State"→"From New Blend Tree"命令。

（3）双击混合树（Blend Tree）以进入混合树视图（Blend Tree Graph）。

Animator 窗口现在显示整个混合树的图示，而 Inspector 显示当前选定节点及其直接子节点（图 14-11）。

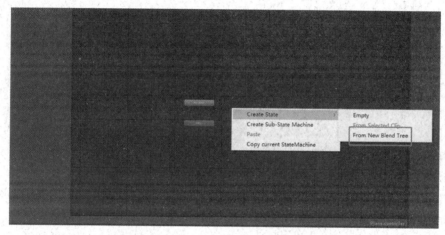

图 14-11

要将动画剪辑添加到混合树，可选择该混合树，然后单击 Inspector 的 Motion 字段内的加号图标（图 14-12）。

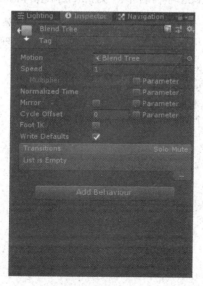

图 14-12

项目15 IK反向运动学

15.1 项目表单

表 15-1 学习性工作任务单

学习场 K	IK 反向运动学				
学习情境 L	了解 IK 反向运动学				
学习任务 M	创建一个关于 IK 反向运动学的项目		学时	4 学时(180 min)	
典型工作过程描述	创建项目—了解 IK 反向运动学—创建一个关于 IK 反向运动学的项目				
学习目标	1. 了解 IK 反向运动学 2. 熟悉 IK 反向运动学 3. 创建一个关于 IK 反向运动学的项目				
任务描述	熟悉 IK 反向运动学并创建一个关于 IK 反向运动学的项目				
学时安排	资讯 20 min	计划 10 min	决策 10 min	实施 100 min	检查 20 min	评价 20 min
对学生的要求	1. 安装好软件 2. 课前做好预习 3. 熟悉 IK 反向运动学 4. 创建一个关于 IK 反向运动学的项目				
参考资料	1. 素材包 2. 微视频 3. PPT				

表 15-2　资讯单

学习场 K	IK 反向运动学		
学习情境 L	了解 IK 反向运动学		
学习任务 M	创建一个关于 IK 反向运动学的项目	学时	20 min
典型工作过程描述	创建项目—了解 IK 反向运动学—创建一个关于 IK 反向运动学的项目		
搜集资讯的方式	1. 教师讲解 2. 互联网查询 3. 同学交流		
资讯描述	查看教师提供的资料，获取信息，便于创建		
对学生的要求	1. 软件安装完成 2. 课前做好预习 3. 熟悉 IK 反向运动学 4. 创建一个关于 IK 反向运动学的项目		
参考资料	1. 素材包 2. 微视频 3. PPT		

表 15-3　计划单

学习场 K	IK 反向运动学		
学习情境 L	了解 IK 反向运动学		
学习任务 M	创建一个关于 IK 反向运动学的项目	学时	10 min
典型工作过程描述	创建项目—了解 IK 反向运动学—创建一个关于 IK 反向运动学的项目		
计划制订的方式	同学间分组讨论		
序号	工作步骤	注意事项	
1	创建一个新的项目		
2	熟悉 IK 反向运动学		
3	创建一个关于 IK 反向运动学的项目		
计划评价	班级		第　　组	组长签字	
	教师签字		日期		
	评语：				

表 15-4 决策单

学习场 K	IK 反向运动学				
学习情境 L	了解 IK 反向运动学				
学习任务 M	创建一个关于 IK 反向运动学的项目			学时	10 min
典型工作过程描述	创建项目—了解 IK 反向运动学—创建一个关于 IK 反向运动学的项目				
计划对比					
序号	计划的可行性	计划的经济性	计划的可操作性	计划的实施难度	综合评价
1					
2					
3					
4					
5					
6					
7					
8					
9					
10					
决策评价	班级		第____组	组长签字	
	教师签字		日期		
	评语:				

表 15-5 实施单

学习场 K	IK 反向运动学		
学习情境 L	了解 IK 反向运动学		
学习任务 M	创建一个关于 IK 反向运动学的项目	学时	100 min
典型工作过程描述	创建项目—了解 IK 反向运动学—创建一个关于 IK 反向运动学的项目		
序号	实施步骤	注意事项	
1	创建一个新的项目	新建项目	
2	了解 IK 反向运动学	IK 反向运动学	
3	创建一个关于 IK 反向运动学的项目	是否合理,是否正确	

实施说明:

1. 启动 UnityHub 程序后,需要创建一个项目
2. 熟悉 IK 反向运动学
3. 创建一个关于 IK 反向运动学的项目

实施评价	班级		第____组	组长签字	
	教师签字		日期		
	评语:				

表 15-6　检查单

学习场 K	IK 反向运动学			
学习情境 L	了解 IK 反向运动学			
学习任务 M	创建一个关于 IK 反向运动学的项目		学时	20 min
典型工作过程描述	创建项目—了解 IK 反向运动学—创建一个关于 IK 反向运动学的项目			
序号	检查项目	检查标准	学生自查	教师检查
1	资讯环节	获取相关信息情况		
2	计划环节	熟悉 IK 反向运动学		
3	实施环节	创建一个关于 IK 反向运动学的项目		
4	检查环节	各个环节逐一检查		
检查评价	班级		第＿＿＿组	组长签字	
	教师签字		日期		
	评语：				

表 15-7　评价单

学习场 K	IK 反向运动学			
学习情境 L	了解 IK 反向运动学			
学习任务 M	创建一个关于 IK 反向运动学的项目		学时	20 min
典型工作过程描述	创建项目—了解 IK 反向运动学—创建一个关于 IK 反向运动学的项目			
评价项目	评价子项目	学生自评	组内评价	教师评价
资讯环节	1. 听取教师讲解 2. 互联网查询情况 3. 同学交流情况			
计划环节	1. 查询资料情况 2. 熟悉 IK 反向运动学			
实施环节	1. 学习态度 2. 对 IK 反向运动学的掌握程度			
最终结果	综合情况			
评价	班级		第＿＿＿组	组长签字	
	教师签字		日期		
	评语：				

表 15-8 教学引导文设计单

学习场 K	IK 反向运动学	学习情境 L	了解 IK 反向运动学	参照系		信息工程学院
		学习任务 M	创建一个关于 IK 反向运动学的项目			
普适性 工作过程 典型工 作过程	资讯	计划	决策	实施	检查	评价
分析 IK 反向 运动学	教师讲解	同学分组讨论	计划的可行性	了解 IK 反 向运动学	获取相关 信息情况	评价学习态度
熟悉 IK 反向运动学	自行掌握	熟悉 IK 反向运动学	计划的经济性	设置操作方式	检查组件	评价学生 的熟悉度
进行运用	创建 IK 反向运动学	设计操作	计划的 实施难度	运用所学 所有知识	检查语句 运用是否正确	软件熟练 程度
保存项目文件	了解项目文件 的格式	了解项目 文件的格式	综合评价	保存项目文件	检查项目 文件的格式	评价项目

表 15-9 教学反馈单(学生反馈)

学习场 K	IK 反向运动学			
学习情境 L	了解 IK 反向运动学			
学习任务 M	创建一个关于 IK 反向运动学的项目		学时	4 学时(180 min)
典型工作过程描述	创建项目—了解 IK 反向运动学—创建一个关于 IK 反向运动学的项目			
调查项目	序号	调查内容	理由描述	
	1	资讯环节		
	2	计划环节		
	3	实施环节		
	4	检查环节		
您对本次课程教学的改进意见:				
调查信息	被调查人姓名		调查日期	

表 15-10　分组单

学习场 K	IK 反向运动学			
学习情境 L	了解 IK 反向运动学			
学习任务 M	创建一个关于 IK 反向运动学的项目		学时	4 学时(180 min)
典型工作过程描述	创建项目—了解 IK 反向运动学—创建一个关于 IK 反向运动学的项目			
分组情况	组别	组长	组员	
	1			
	2			
	3			
	4			
	5			
	6			
	7			
	8			
分组说明				
班级		教师签字		日期

表 15-11　教师实施计划单

学习场 K	IK 反向运动学					
学习情境 L	了解 IK 反向运动学					
学习任务 M	创建一个关于 IK 反向运动学的项目		学时	4 学时(180 min)		
典型工作过程描述	创建项目—了解 IK 反向运动学—创建一个关于 IK 反向运动学的项目					
序号	工作与学习步骤	学时	使用工具	地点	方式	备注
1	资讯情况	20 min	互联网			
2	计划情况	10 min	计算机			
3	决策情况	10 min	计算机			
4	实施情况	100 min	Unity			
5	检查情况	20 min	计算机			
6	评价情况	20 min	课程伴侣			
班级		教师签字		日期		

表 15-12　成绩报告单

班级IK 反向运动学学习场(课程)成绩报告单														
学习场 K	IK 反向运动学													
学习情境 M	了解 IK 反向运动学													
典型工作过程描述	创建项目—了解 IK 反向运动学—创建一个关于 IK 反向运动学的项目								学时			4 学时(180 min)		
序号	姓名	第一个学习任务				第二个学习任务				第 N 个学习任务				总评
		自评 ×%	互评 ×%	教师评 ×%	合计	自评 ×%	互评 ×%	教师评 ×%	合计	自评 ×%	互评 ×%	教师评 ×%	合计	
1														
2														
3														
4														
5														
6														
7														
8														
9														
10														
11														
12														
13														
14														
15														
16														
17														
18														
19														
20														
21														
22														
23														
24														
25														
26														
27														
28														
班级		教师签字								日期				

15.2 理论指导

大多数动画是通过将骨架中的关节角度旋转到预定值来生成的。子关节的位置根据父关节的旋转而改变，因此可从父关节包含的各个关节的角度和相对位置来确定关节链的终点。这种构建骨架的方法被称为正向动力学。

然而，从相反视角看待构建关节的任务通常很有用：在空间中选择一个位置后，向后找到一种有效的关节定位方法，使终点落在该位置。如果想要角色触摸位于用户选定位置的对象或想要角色的双脚牢牢扎入不平坦的表面，这种方法可能很有用。此方法称为反向动力学(IK)，可在 Mecanim 中用于已正确配置的任何人形 Avatar 骨骼。

要为角色设置 IK，通常要在场景周围放置与角色互动的对象，然后通过脚本(尤其是 SetIKPositionWeight、SetIKRotationWeight、SetIKPosition、SetIKRotation、SetLookAt-Position、bodyPosition、bodyRotation 之类的 Animator 函数)来设置 IK。

(1)SetIKPositionWeight(图 15-1)。设置反向动力学目标的转换权重(0＝在反向动力学前的原始动画处，1＝在目标处)。反向动力学目标是特定身体部位的目标位置和旋转。Unity 能够计算如何将身体部位从起点(从动画中获得的当前位置和旋转)向目标移动。该函数设置 0~1 范围内的权重值，以确定起始位置与反向动力学将瞄准的目标位置之间的距离。使用 SetIKPosition 单独设置位置本身。

Animator.SetIKPositionWeight

public void SetIKPositionWeight (AvatarIKGoal goal, float value);

参数

| goal | 动画的 AvatarIKGoal。 |
| value | 转换权重。 |

图 15-1

(2)SetIKRotationWeight(图 15-2)。设置反向动力学目标的旋转权重(0＝在反向动力学前旋转，1＝在反向动力学目标处旋转)。反向动力学目标是特定身体部位的目标位置和旋转。Unity 能够计算如何将身体部位从起点(从动画中获得的当前位置和旋转)向目标移动。该函数设置 0~1 范围内的权重值，以确定起始旋转与反向动力学将瞄准的目标旋转之间的距离。使用 SetIKRotation 单独设置目标本身。

Animator.SetIKRotationWeight

public void SetIKRotationWeight (AvatarIKGoal goal, float value);

参数

| goal | 记到的 AvatarIKGoal。 |
| value | 旋转权重。 |

图 15-2

（3）SetIKPosition（图 15-3）。设置反向动力学目标的位置。反向动力学目标是特定身体部位的目标位置和旋转。Unity 能够计算如何将身体部位从起点（从动画中获得的当前位置和旋转）向目标移动。该函数设置最终目标在世界空间中的位置；身体部位最终在空间中的实际点也受到权重参数的影响，权重参数指定起点与反向动力学应瞄准的目标之间的距离（范围 0～1 的值）。应始终在 MonoBehaviour. OnAnimatorIK 中调用该函数。

图 15-3

（4）SetIKRotation（图 15-4）。设置反向动力学目标的旋转。反向动力学目标是特定身体部位的目标位置和旋转。Unity 能够计算如何将身体部位从起点（从动画中获得的当前位置和旋转）向目标移动。该函数设置最终目标在世界空间中的旋转；身体部位最终的实际旋转也受到权重参数的影响，权重参数指定起点与反向动力学应瞄准目标之间的距离（范围 0～1 的值）。

图 15-4

项目16 XML文件

16.1 项目表单

表 16-1 学习性工作任务单

学习场 K	XML 文件					
学习情境 L	了解 XML 文件					
学习任务 M	创建 XML 文件			学时	4 学时(180 min)	
典型工作过程描述	创建项目—了解 XML 文件—创建 XML 文件					
学习目标	1. 了解 XML 文件 2. 熟悉 XML 文件 3. 创建 XML 文件					
任务描述	熟悉 XML 文件并创建 XML 文件					
学时安排	资讯 20 min	计划 10 min	决策 10 min	实施 100 min	检查 20 min	评价 20 min
对学生的要求	1. 安装好软件 2. 课前做好预习 3. 熟悉 XML 文件 4. 创建 XML 文件					
参考资料	1. 素材包 2. 微视频 3. PPT					

表 16-2　资讯单

学习场 K	XML 文件		
学习情境 L	了解 XML 文件		
学习任务 M	创建 XML 文件	学时	20 min
典型工作过程描述	创建项目—了解 XML 文件—创建 XML 文件		
搜集资讯的方式	1. 教师讲解 2. 互联网查询 3. 同学交流		
资讯描述	查看教师提供的资料，获取信息，便于创建		
对学生的要求	1. 软件安装完成 2. 课前做好预习 3. 熟悉 XML 文件 4. 创建 XML 文件		
参考资料	1. 素材包 2. 微视频 3. PPT		

表 16-3　计划单

学习场 K	XML 文件			
学习情境 L	了解 XML 文件			
学习任务 M	创建 XML 文件		学时	10 min
典型工作过程描述	创建项目—了解 XML 文件—创建 XML 文件			
计划制订的方式	同学间分组讨论			
序号	工作步骤		注意事项	
1	创建一个新的项目			
2	熟悉 XML 文件			
3	创建 XML 文件			
计划评价	班级		第___组	组长签字
	教师签字		日期	
	评语：			

表 16-4 决策单

学习场 K	XML 文件		
学习情境 L	了解 XML 文件		
学习任务 M	创建 XML 文件	学时	10 min
典型工作过程描述	创建项目—了解 XML 文件—创建 XML 文件		

计划对比

序号	计划的可行性	计划的经济性	计划的可操作性	计划的实施难度	综合评价
1					
2					
3					
4					
5					
6					
7					
8					
9					
10					

决策评价	班级		第____组	组长签字	
	教师签字		日期		
	评语：				

表 16-5 实施单

学习场 K	XML 文件		
学习情境 L	了解 XML 文件		
学习任务 M	创建 XML 文件	学时	100 min
典型工作过程描述	创建项目—了解 XML 文件—创建 XML 文件		

序号	实施步骤	注意事项
1	创建一个新的项目	新建项目
2	了解 XML 文件	XML 文件
3	创建 XML 文件	是否合理，是否正确

实施说明：
1. 启动 UnityHub 程序后，需要创建一个项目
2. 熟悉 XML 文件
3. 创建 XML 文件

实施评价	班级		第____组	组长签字	
	教师签字		日期		
	评语：				

表 16-6   检查单

学习场 K	XML 文件			
学习情境 L	了解 XML 文件			
学习任务 M	创建 XML 文件		学时	20 min
典型工作过程描述	创建项目—了解 XML 文件—创建 XML 文件			
序号	检查项目	检查标准	学生自查	教师检查
1	资讯环节	获取相关信息情况		
2	计划环节	熟悉 XML 文件		
3	实施环节	创建 XML 文件		
4	检查环节	各个环节逐一检查		
检查评价	班级		第___组	组长签字	
	教师签字		日期		
	评语：				

表 16-7   评价单

学习场 K	XML 文件			
学习情境 L	了解 XML 文件			
学习任务 M	创建 XML 文件		学时	20 min
典型工作过程描述	创建项目—了解 XML 文件—创建 XML 文件			
评价项目	评价子项目	学生自评	组内评价	教师评价
资讯环节	1. 听取教师讲解 2. 互联网查询情况 3. 同学交流情况			
计划环节	1. 查询资料情况 2. 熟悉 XML 文件			
实施环节	1. 学习态度 2. 对 XML 文件的掌握程度			
最终结果	综合情况			
评价	班级		第___组	组长签字	
	教师签字		日期		
	评语：				

表 16-8　教学引导文设计单

学习场 K	XML 文件	学习情境 L	了解 XML 文件	参照系	信息工程学院
		学习任务 M	创建 XML 文件		
普适性工作过程 / 典型工作过程	资讯	计划	决策	实施	检查	评价
分析 XML 文件	教师讲解	同学分组讨论	计划的可行性	了解 XML 文件	获取相关信息情况	评价学习态度
熟悉 XML 文件	自行掌握	熟悉 XML 文件	计划的经济性	设置操作方式	检查组件	评价学生的熟悉度
进行运用	创建 XML 文件	设计操作	计划的实施难度	运用所学所有知识	检查语句运用是否正确	软件熟练程度
保存项目文件	了解项目文件的格式	了解项目文件的格式	综合评价	保存项目文件	检查项目文件的格式	评价项目

表 16-9　教学反馈单(学生反馈)

学习场 K	XML 文件		
学习情境 L	了解 XML 文件		
学习任务 M	创建 XML 文件	学时	4 学时(180 min)
典型工作过程描述	创建项目—了解 XML 文件—创建 XML 文件		
调查项目	序号	调查内容	理由描述
	1	资讯环节	
	2	计划环节	
	3	实施环节	
	4	检查环节	

您对本次课程教学的改进意见:

| 调查信息 | 被调查人姓名 | | 调查日期 | |

表 16-10　分组单

学习场 K	XML 文件			
学习情境 L	了解 XML 文件			
学习任务 M	创建 XML 文件		学时	4 学时(180 min)
典型工作过程描述	创建项目—了解 XML 文件—创建 XML 文件			
分组情况	组别	组长	组员	
	1			
	2			
	3			
	4			
	5			
	6			
	7			
	8			
分组说明				
班级		教师签字	日期	

表 16-11　教师实施计划单

学习场 K	XML 文件					
学习情境 L	了解 XML 文件					
学习任务 M	创建 XML 文件		学时	4 学时(180 min)		
典型工作过程描述	创建项目—了解 XML 文件—创建 XML 文件					
序号	工作与学习步骤	学时	使用工具	地点	方式	备注
1	资讯情况	20 min	互联网			
2	计划情况	10 min	计算机			
3	决策情况	10 min	计算机			
4	实施情况	100 min	Unity			
5	检查情况	20 min	计算机			
6	评价情况	20 min	课程伴侣			
班级		教师签字		日期		

表 16-12 成绩报告单

		_____班级XML文件学习场(课程)成绩报告单													
学习场 K		XML 文件													
学习情境 M		了解 XML 文件													
典型工作过程描述		创建项目—了解 XML 文件—创建 XML 文件								学时		4 学时(180 min)			
序号	姓名	第一个学习任务				第二个学习任务				第 N 个学习任务				总评	
		自评 ×%	互评 ×%	教师评 ×%	合计	自评 ×%	互评 ×%	教师评 ×%	合计	自评 ×%	互评 ×%	教师评 ×%	合计		
1															
2															
3															
4															
5															
6															
7															
8															
9															
10															
11															
12															
13															
14															
15															
16															
17															
18															
19															
20															
21															
22															
23															
24															
25															
26															
27															
28															
班级			教师签字						日期						

16. 2　理论指导

XML 意为可扩展标记语言，用来存取数据和存放相应的游戏配置，是一种应用程序之间数据传输的常用格式。XML 文件相比于 JSON 而言，稍微复杂一点。而且现在基本上都倾向于用 JSON，更加轻量级。

1. XML 文件的格式

XML 文件的格式如下：

＜? xml version＝"1. 0" encoding＝"UTF-8"? ＞

＜SceneName＞

＜ObjectName＞DataSaver＜/ObjectName＞

＜/SceneName＞

第一行为 XML 的声明，它的作用是将文件标记为 XML 文件，有利于快速被识别，所以这个声明必须放在文件的最顶部才有效。在创建 XML 元素时，名称可以使用英文字母、数字、特殊字符，比如下画线。除此之外，还要注意以下几个方面：

(1)元素名中不可以有空格出现。

(2)大小写不限制，但最好整个文件进行统一。

(3)元素名只能以英文字母开始，不可以是数字或者符号。

2. 创建 XML 文件

(1)加入两个命名空间：

①对 XML 文件的命名空间，里面有关 XML 的 APL 很多，using System. Xml；

②对文件读写的命名空间，就是对文件的读取，保存操作，using System. IO；

代码示例如图 16-1 所示。

(2)结果如图 16-2 所示。

(3)增删查改如图 16-3 所示。

```
void CreateXml()
{
    string path = Application.dataPath+"/item.xml";

    XmlDocument xml = new XmlDocument();
    XmlElement root = xml.CreateElement("参考资料");
    XmlElement element = xml.CreateElement("书籍");
    element.SetAttribute("id","1");
    XmlElement elementchild1 = xml.CreateElement("名称");
    elementchild1.InnerText = "xml精讲";
    XmlElement.elementchild2 = xml.CreateElement("价格");
    elementchild2.SetAttribute("货币单位","人民币");
    elementchild2.InnerText = "18.00";
    element.AppendChild(elementchild1);
    element.AppendChild(elementchild2);

    root.AppendChild(element);
    xml.AppendChild(root);
    xml.Save(path);
}
```

图 16-1

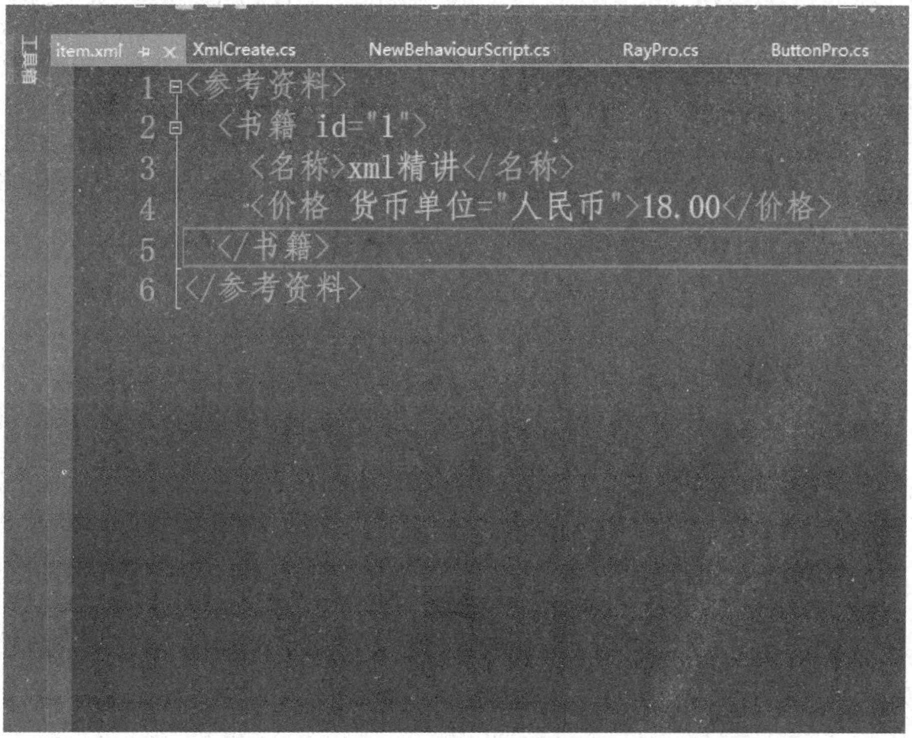

item.xml | XmlCreate.cs | NewBehaviourScript.cs | RayPro.cs | ButtonPro.cs

```
1 <参考资料>
2    <书籍 id="1">
3        <名称>xml精讲</名称>
4        <价格 货币单位="人民币">18.00</价格>
5    </书籍>
6 </参考资料>
```

图 16-2

```
43      void SelectXml()
44      {
45          string path = Application.dataPath + "/item.xml";
46          if (File.Exists(path))
47          {
48              XmlDocument xml = new XmlDocument();
49              xml.Load(path);
50              XmlNodeList xmlNodeList = xml.SelectSingleNode("参考资料").ChildNodes;
51              foreach (XmlElement item in xmlNodeList)
52              {
53                  foreach (XmlElement itemchild in item.ChildNodes)
54                  {
55                      //获取值删除节点
56                      if (itemchild.Name == "价格")
57                      {
58
59
60                          item.RemoveChild(itemchild);
61                      }
62                      //设置属性 删价属性
63                      if(itemchild.Name =="名称")
64                      {
65                          Debug.Log("ming");
66                          itemchild.SetAttribute("name", "123");
67                          itemchild.RemoveAttribute("name");
68                          //XmlAttribute xmlAttri = xml.CreateAttribute("name");
69                          //xmlAttri.Value = "123";
70                          //itemchild.Attributes.Append(xmlAttri);
71                      }
72                  }
73              }
74
75              xml.Save(path);
76          }
77      }
78  }
```

图 16-3

Project 1　Unity Basic Interface

1. 2　Theoretical guidance

1. 2. 1　Unity Basic Interface layout

When we open the Unity editor, can see many views, including Hierarchy, Scene, Project, Inspector and Game, which are very closely related to each other and can see the whole game project hierarchy, structure and concept more clearly(Figure 1-1).

Figure 1-1

1. Hierarchy view

The Hierarchy view is mainly used to store the specific game objects and the hierarchical relationships between them in the game scene. Parent-child relationships between game objects can be created in the Hierarchy view(Figure 1-2).

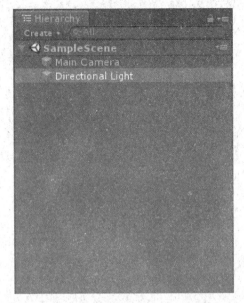

Figure 1-2

2. Scene view

The Scene view is used to edit the entire game world, including the various art re-sources and the system's own resources (Gimmes), and to build them in the Scene view, including lighting, models, effects, etc. Select the cube object in Hierarchy view and press the shortcut F in Scene view to see the game object up close(Figure 1-3).

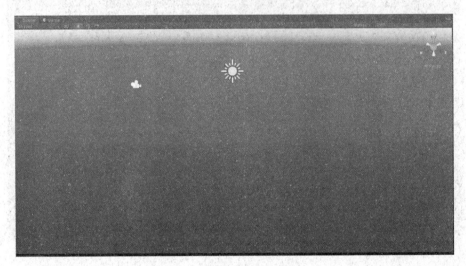

Figure 1-3

3. Project view

This is used to store all the resource files and scripts used in the design of the game; in the top left corner of the Project view there is a Create option where you can create a new game resource by clicking on it(Figure 1-4).

4. Inspector view

A place for presenting game objects, game resources, game settings and for displaying description information and properties. In this view, there is a description of the selected component and all parameters of that component's description, and some of the component parameters can be dynamically modified(Figure 1-5).

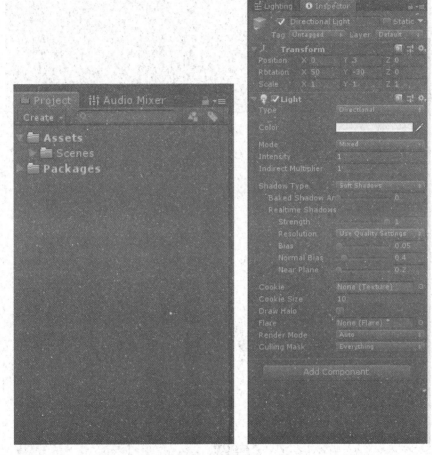

Figure 1-4　　　　　　　　　　　　　　　　Figure 1-5

5. Game view

The final game that will be displayed on the screen when it is released is the result of the game being run. The screen display depends entirely on the part of the Hierarchy view that is illuminated by the camera(Figure 1-6).

1.2.2　Top menu bar

As with other software, the top menu bar contains Unity's main functions and settings. There are seven menu bars as shown.

Figure 1-6

1. The File menu

The File menu is mainly used to create and store scenes and projects(Figure 1-7).

2. The Edit menu

The Edit menu is mainly used for editing the settings of the individual objects in the scene(Figure 1-8).

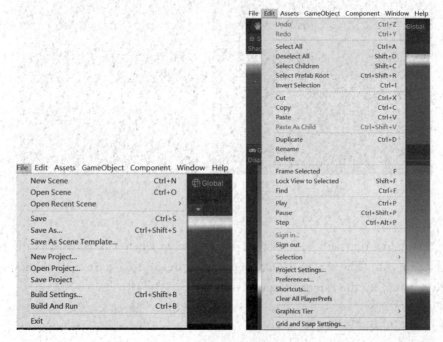

Figure 1-7   Figure 1-8

3. The Assets menu

The Assets menu allows you to import or export the resource packages used, and is provided by Unity to manage the resources needed for the game(Figure 1-9).

4. The Game Object menu

The Game Object menu is mainly used to add game objects to the scene and to set up the objects(Figure 1-10).

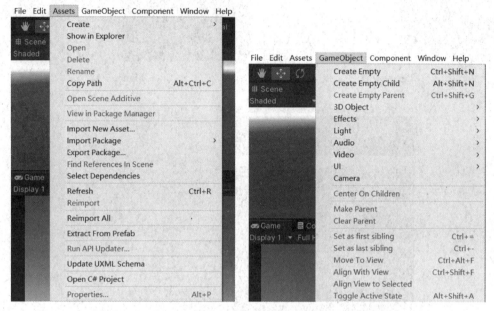

Figure 1-9   Figure 1-10

5. The Component menu

The Component menu is mainly used to add components or properties to game objects during the creation of a project(Figure 1-11).

6. The Window menu

The Window menu controls the page layout of the entire editor and controls the opening and closing of individual view windows(Figure 1-12).

7. Help menu

The Help menu is used to help users learn and master Unity3D(Figure 1-13).

1. 2. 3   Toolbar

The Toolbar contains seven basic controls. Each control is associated with a different part of the Editor.

1. Transform component tool

Used to manipulate the Scene view, mainly to enable control of game objects, including position, rotation, scaling, etc(Figure 1-14).

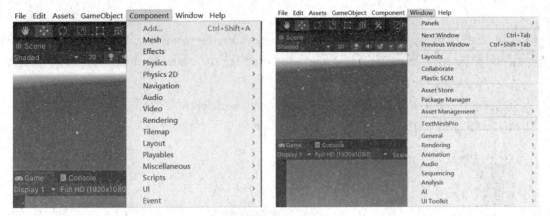

Figure 1-11   Figure 1-12

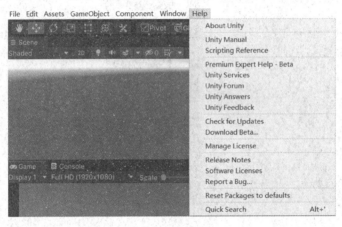

Figure 1-13

2. Change auxiliary icon switch

Affects the display of the Scene view and mainly transforms the position of game objects(Figure 1-15).

Figure 1-14   Figure 1-15

3. Play/Pause/Step buttons

Used to handle the Game view, mainly for controlling the project, playing, pausing and stepping through the project(Figure 1-16).

Figure 1-16

4. Cloud button

Opens the Unity Services window(Figure 1-17).

5. Account drop-down menu

Used to access your Unity account(Figure 1-18).

Figure 1-17　　　　**Figure 1-18**

6. Layers drop-down menu

Controls which objects are displayed in the Scene view(Figure 1-19).

7. Layout drop-down menu

Controls the layout of all views(Figure 1-20).

Figure 1-19　　　　**Figure 1-20**

Project 2 Unity Basic interface operation

2.2 Theoretical guidance

Unity basic interface operations, there are a few points.

1. Scene view manipulation

The Scene view has a set of navigation controls that can be used to move quickly and efficiently.

2. Scene view helper icons

The Scene View Assist icon (Scene Gizmo) is located in the top right corner of the Scene view. This control is used to display the current orientation of the Scene view camera and allows quick modification of the view and projection mode(Figure 2-1).

3. Flight mode

(1) Click and hold down the right mouse button.

(2) Use the mouse to move the view, use the W/A/S/D keys to move left/right/forward/backward, and use the Q and E keys to move up and down.

Figure 2-1

(3) Press and hold the Shift key to move faster.

4. moving, rotating, scaling, rectangular transform, transformation game objects

The first tool in the toolbar, the Hand Tool, is used for panning the scene and has the shortcut Q. Selecting the Hand Tool and holding Alt rotates the current view of the scene. Alternatively, holding Alt and dragging left or right with the right mouse button will zoom in and out of the scene, the same effect can be achieved with the mouse wheel. The Move, Rotate, Scale, Cert Transform and Transform tools are used to edit individual game objects. To change the transform component of a game object, use the mouse to manipulate any of the secondary icon axes or enter a value directly into the numeric field of the transform component in the Inspector. You can also use the shortcut keys to select transformation modes: W for

move, E for rotate, R for scale, T for rectangle
transformation and Y for transform(Figure 2-2).

(1)Moving. In the centre of the Move Assist icon,

Figure 2-2

there are three small squares that can be used to drag
game objects in a single plane (meaning that two axes can be moved at once while the third
remains stationary), as well as three axes, representing the ability to move to any axis
(Figure 2-3).

(2)Rotation. Once you have selected the Rotate tool, change the rotation of the game
object by clicking and dragging the axis of the wire frame sphere auxiliary icon displayed a-
round the game object(Figure 2-4).

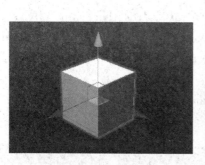

Figure 2-3   Figure 2-4

(3)Scaling. Use the zoom tool to re-scale the game object evenly on all axes by clicking
and dragging the cube in the centre of the auxiliary icon(Figure 2-5).

(4)Rectangular Transformations. Rectangle transforms are usually used to position 2D
elements such as sprites or UI elements, but can also be used to manipulate 3D game objects. This tool combines the move, scale and rotate functions into one helper icon:

①Click and drag a rectangular helper icon to move a game object.

②Click and drag any corner or edge of the rectangular helper icon to scale the game
object.

③Drag an edge to scale the object along one axis.

④Drag a corner to scale the game object on both axes.

⑤To rotate the game object, place the cursor outside one of the corners of the rectangle. The cursor changes to show the rotation icon. Click and drag from this area to rotate
the game object(Figure 2-6).

(5)Transformation. The Transform tool combines the Move, Rotate and Scale tools.
The tool's secondary icon provides handles for movement and rotation(Figure 2-7).

5. Parent and child objects

(1)A game object can have only one parent object.

(2)A game object can have numerous child objects.

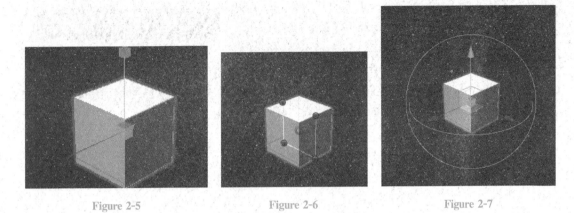

Figure 2-5   Figure 2-6   Figure 2-7

(3)When a child object is shifted, rotated or scaled, there is no effect on the parent object. The reason: the position, rotation and scaling of a child object is relative to the parent object and the position relative to the parent object remains the same regardless of how the parent object is moved.

(4)When a parent object is shifted, rotated or scaled, it has an effect on all the child objects(Figure 2-8).

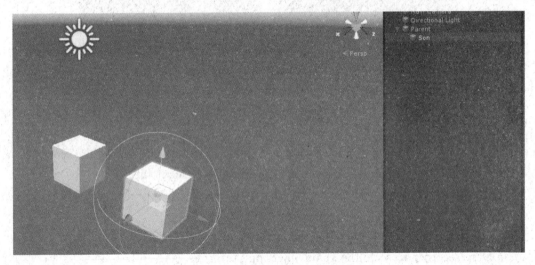

Figure 2-8

Project 3　Inspector panel function, Camera component,Light component function introduction

3.2　Theoretical guidance

Inspector panel function，Camera component，Light component function introduction is as follows.

3.2.1　Inspector view

A project in the Unity Editor consists of multiple game objects that contain scripts, sounds，meshes and other graphical elements（such as light sources）The Inspector window（sometimes called Inspector）displays detailed information about the currently selected game object，including all additional components and their properties，and allows the functionality of the game object in the scene to be modified(Figure 3-1).

1. Inspecting game objects and script variables

Use Inspector to view and edit the properties and settings of almost everything in the Unity Editor，including physical game elements such as game objects，resources and materials，as well as the settings and preferences within the Editor. When a game object is selected in Hierarchy or Scene view，the Inspector will display the properties of all components and materials for that game object,use the Inspector to edit the settings of these components and materials(Figure 3-2).

Figure 3-1

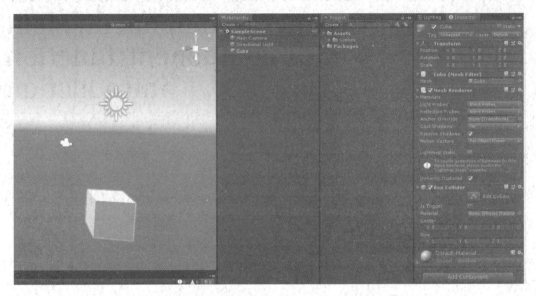

Figure 3-2

2. Inspecting resources

After selecting a resource in the Project window, Inspector will display settings on how to import the resource and use it at runtime (when the game is running in the Editor or in a published version). Each type of resource has a different set of settings. For example the material resources are shown in the figure(Figure 3-3).

3. Game Object Icons (Select Icon)

After assigning an icon to a game object, the icon will be displayed above that game object (and any subsequent repeating items) in the Scene view. Icons can also be assigned to a prefabricated item, thus applying the icon to all instances of that prefabricated item in the scene(Figure 3-4).

4. Showing and hiding game objects

In Unity, created a Game Object to be placed on the Hierarchy view. If you want to hide the object, can set it via the Inspector view, with the option at the top, uncheck it to hide the object(Figure 3-5).

5. Setting static

Objects can be set to static via the top right corner(Figure 3-6).

6. Setting Tags and Layers

Tags and Layers are the properties used to identify Game Objects in the Unity engine, Tags are commonly used for a single Game Object and Layers are commonly used for a group of Game Objects.

Execute "Edit"→"Project Settings"→"Tags and Layers" command to open the settings panel. You can also add them directly from the drop down menu(Figure 3-7).

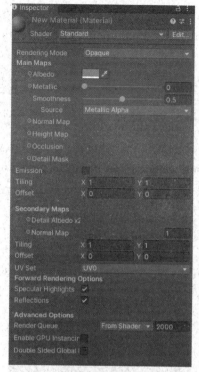

Figure 3-3

Figure 3-4

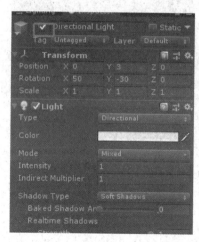

Figure 3-5

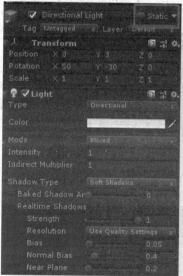

Figure 3-6

Figure 3-7

7. Introduction to components

A Game Object contains several components. A component represents a function (Figure 3-8).

3.2.2　Camera component features

The Camera is a device that captures and displays the world for the player. By customizing and manipulating the camera, can give your game a truly unique character. Can have an unlimited number of cameras in the scene. These cameras can be set to render anywhere on the screen in any order or only in certain parts of the screen(Figure 3-9).

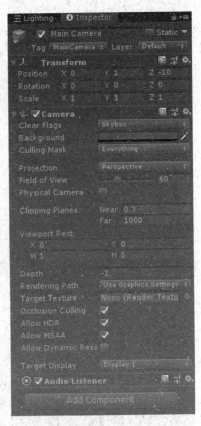

Figure 3-8　　　　　　　　　　Figure 3-9

Introduction to Camera properties:

(1) Clear Flags: Determines which parts of the screen will be cleared. This is handy when using multiple cameras to draw different game elements.

(2) Background: The colour to be applied to the remaining part of the screen after all elements in the view have been drawn but without the skybox.

(3) Culling Mask: elects the Layer to be displayed.

(4) Projection:

①Perspective: The camera will render the game object in perspective.

②Field of view: Used to control the width of the camera's field of view and the angular size of the portrait.

③Orthographic: The camera will render the game object without perspective.

④Size: Controls the size of the viewport of the camera in orthogonal mode.

(5) Clipping Planes: The distance between when the camera starts rendering and when it stops rendering.

①Near: The closest point. The closest point at which the camera starts rendering.

②Far: Far point. The farthest point from where the camera starts rendering.

(6) Viewport Cert: Four values are used to control the position and size of the camera 's view that will be drawn on the screen, using a screen coordinate system with values between 0~1. The origin of the coordinate system is in the bottom left corner.

X: position from the x-axis;

Y: position from the y-axis;

W: Weight the width of the image rendered by the game object;

H: Height the height of the image rendered by the game object.

(7)Depth: Controls the order in which the cameras are rendered, with the larger camera being rendered on top of the smaller one.

(8) Target Texture: The target image. You can freely set the background image under the game form.

3. 2. 3　Introduction to the Light component functionality

Light in Unity is mainly provided by the Light object. Light sources determine the colouring of objects and the shadows they cast. As such, they are a fundamental part of graphics rendering(Figure 3-10).

1. Type Light Type

The possible values are Directional Light, Point Light , Spot Light and Area Light v (area light, which is only reflected under baking).

(1) Range. Defines the distance that the light from the centre of the object travels (Point and Spot only).

(2) Spot Angle. Defines the angle (in degrees) at the base of the spotlight cone (Spot only).

2. Color

Use the colour picker to set the colour of the light.

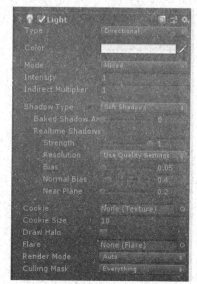

Figure 3-10

3. Mode

Specify the lighting mode, which determines if and how the light source is "baked". Modes may be Real time, Mixed (between Real time and Baked) and Baked.

4. Intensity

Sets the brightness of the light source. The default value for Direction is 0.5, and for Point, Spot or Area the default value is 1.

(1)Indirect Multiplier. Use this value to change the intensity of the indirect light. Indirect light is light that is bounced from one object to another. Indirect Multiplier defines the brightness of the scattered light calculated by the Global Illumination (GI) system. If the Indirect Multiplier is set to a value below 1, each bounce will make the scattered light dimmer. A value greater than 1 makes the light brighter after each bounce. This is useful, for example, to brighten the shadowy side of a shaded area (e. g. the inside of a cave) to the point where it is clearly visible. If you want to use live global lighting but want to limit a single live light source so that it only emits direct light, set its Indirect Multiplier to 0.

(2)Shadow Type. Decide whether this light source casts hard shadows, soft shadows or no shadows at all. Please see the documentation on shadows for information on hard and soft shadows.

Project 4　How to write scripts in Unity

4.2　Theoretical guidance

Definition and use of C# variables introduction is as follows.

1. Basic syntax and use of C#

(1) Definition of C# variables. In a programming language, every value needs to be stored in a particular place, which call a variable. A variable consists of a variable name and a data type. The data type of the variable determines which type of data can store in it.

When defining a variable, you first need to identify the data type of the value to be stored in the variable, then determine the contents of the variable and finally define the variable name according to the C# variable naming rules.

The syntax for defining variables is as follows.

Data type Variable name.

For example, to define a variable that holds an integer, it could be defined as int num;

Assign values to variables after they are defined, just use "=" to concatenate the value to be stored in the variable.

(2) The following data types are common:

① Numeric variables. In C#, numeric variables are mainly integers (int), single precision floating point numbers (float) and double precision floating point numbers (double). float and double variables differ in that float has 6 valid bits and takes up 4 bytes of storage space. Whereas double has 15 valid bits and takes up 8 bytes of storage space. By default, decimals are represented by double. If you need to use a float, you need to add f to the end. For example: float a=1.23f; In the above code, defined a floating-point variable called a with an initial value of 1.23.

In addition to the above common types of numeric variables, there are other types of numeric variables available in C#, as shown in Table 4-13.

Table 4-13　the numeric Variables of C#

keywords	instructions	byts size
bool	logical value(true/false)	1
sbyte	signed 8-bit integer	1
byte	unsigned 8-bit integer	1
short	signed 16-bit integer	2
ushort	unsigned 16-bit integer	2
int	signed 32-bit integer	4
uint	unsigned 32-bit integer	4
long	signed 64-bit integer	8
ulong	unsigned 64-bit integer	8
char	16-bit character type	2
float	32-bit single precision floating point type	4
double	64-bit double floating point type	8
decimal	128-bit high precision floating point type	16

They are used in much the same way and will not be repeated.

② Variables of the text type. Variables of the text type are mainly char and string, where char is used to store the value of a single character and string is used to store the value of a string. For example: string a＝"hello";

In the above code, defined a text-based variable named a, of type string (string), whose initial value is hello. it is also possible to save numbers with the string type. string a＝"1"; however, the data stored in a is the string "1", not just the number 1.

③Variables of Boolean type. In C#, a boolean variable is a bool, which is used to store the logical state of a variable, and contains two values: true and false (flase). For example: bool isPlay＝true;In the above code, we have defined a boolean variable called isPlay, with an initial value of true. It is important to note that C# is case-sensitive and case should not be mixed up. For example:Bool isPlay＝true;This statement will report an error in the program because there are no variables of type Bool in C#, only bool.

(3) In addition, C# supports a special type of variable, the constant. Constants are variables that cannot be changed. Constants can only be defined at the class attribute level and must be static and initialised at the time of definition. Constants can only be declared for value types because they are assigned at the time of definition.

To declare a variable, simply add the keyword const or readonly to the front of the variable to specify it as a constant. For example.

```
const int constNum＝2;
readonly int readonlyNum＝3;
```

(4) Rules for naming constants and variables.

① Can only consist of letters, numbers, @ and underscores.

② Cannot start with a number.

③ The @ symbol can only be placed first.

④ Can not be renamed with the system variable keyword.

⑤ No renaming: C# is case-sensitive.

⑥ Chinese variable names are syntactically valid, but it is better not to use them.

(5) Naming conventions for constants and variables.

① Use English words, not pinyin.

② Small hump naming: the first word is not capitalised, the first letter of each subsequent word is capitalised: myHeroDamage.

③ See the name to know the meaning.

2. Assignment of C# variables

There are two ways of assigning syntax, either directly while defining the variable, or by defining the variable first and then assigning it, and they have the following format.

(1)Assigning a value while defining a variable.

Data type Variable name=value.

(2)Defining the variable first and then assigning the value.

Data type Variable name.

Variable name=value.

When defining variables care needs to be taken that the value in the variable is compatible with the data type of the variable. Alternatively,When assigning values to variables it is also possible to assign values to more than one variable at a time. For example. int a=1, b=2;Although it is much easier to assign values to more than one variable at a time, in practice, to enhance the readability of the program,advised to declare and assign values to one variable at a time in the programming.

3. Expressions and Operators

Expressions and operators are used to perform various forms of arithmetic on data or information, and they form the bulk of program code.

Expressions are made up of operators and operands.

Classification and priority of operators(Table 4-14).

In C#, there are several main operators that we need to understand.

① Arithmetic operators. Arithmetic operators are the basic mathematical operations that we are all familiar with. The main ones are add +, subtract -, multiply *, divide /, and find remainder %.

② Assignment operators. The assignment operators are used to assign a piece of data to a variable, property or reference. The data can be a constant, a variable or an expression. The assignment operators themselves are divided into simple and compound assignments, which work as shown in the diagram(Table 4-15).

Table 4-14 Classification and priority of operators

operator classification		operator priority		
class of operator	operators	class	computation sequence	operators
basic arithmatic operation	$+,-,*,/,\%$	basic	high	$x,y,f(x),a[x],x++,x--$
progressive increase and decrease	$++,--$	unitary		$+,-,!,\sim,++x,--x,(T)x$
displacement	$<<,>>$	multiply-divide		$+,/,\%$
logic	$\&,\mid,\wedge,!,-,\&\&,\mid\mid$	addition and substraction		$+,-$
assignment	$=,+=,-=,*=,/=,\%=,\&=,\mid=,\wedge=,<<=,>>=$	displayment		$<<,>>$
relationship	$==,!=,<,>,<=,>=$	relationship		$<,>,<=,>=$
character concatenation	$+$	equality		$==,!=$
member access	.	logic AND		$\&$
index	[]	logic XOR		$\wedge$
transformation	()	logic OR		$\mid$
conditional operation	?:	condition AND		$\&\&$
		condition OR		$\mid$
		condition		?:
		assignment	low	$=,+=,-=,*=,/=,\%=,\&=,\mid=,\wedge=,<,<=,>,>=$

Table 4-15 Assignment operators

assignment operators	sample expressions	meaning
$=$	$x=10$	$x=10$
$+=$	$x+=y$	$x=x+y$
$-=$	$x-=y$	$x=x-y$
$*=$	$x*=y$	$x=x*y$
$/=$	$x/=y$	$x=x/y$
$\%=$	$\%=y$	$x=x\%y$
$>>=$	$x>>=y$	$x=x>>=y$
$<<=$	$x<<=y$	$x=x<<=y$
$\&=$	$x\&=y$	$x=x\&y$
$\mid=$	$x\mid=y$	$x=x\mid y$
$\wedge=$	$x\wedge=y$	$x=x\wedge y$

③ Relational Operators. Relational operators are used to compare two values and return a Boolean result after the comparison. Commonly used relational operators are shown in Table 4-16.

Table 4-16　Commonly used relational operators

relational operators	instructions
==	equal to
<	less than
<=	less than or equal to
>	greater than
>=	greater than or equal to
! =	unequal to

④ Conditional operators. The conditional operators are used to make logical judgements and return a Boolean result. Commonly used conditional operators are shown in Table 4-17.

Table 4-17　Commonly used conditional operators

conditional operators	instructions
&&	and
ll	or
!	non
?:	ternary operator

4. Process control

(1) if.

Syntactic form.

if (expression)

{

statement.

}

(2) if.... else.

Syntactic form.

if (expression)

{

Statement block 1.

}

else

{

statement block 2.

}

(3) while.

Syntactic form.

while(expression)

{

```
    loop body
}
```

(4) do.... while.

Syntax form.

```
do
{
    The loop body statement ;
}
while (conditional expression).
```

(5) for.

syntactic form :

```
for(expression1; expression2).
{
    The loop body statement.
}
```

(6) foreach.

Syntactic form:

```
foreach(type variable name in set object)
{
statement body
}
```

5. Arrays

An array is an ordered collection of fixed size that stores elements of the same type. An array is a collection used to store data and is usually considered to be a collection of variables of the same type.

Declaring arrays. To declare an array in C#, you can use the following syntax.

datatype[ ] arrayName;

where, datatype is used to specify the type of the elements being stored in the array. [ ] specifies the rank (dimension) of the array. The rank specifies the size of the array; arrayName specifies the name of the array.

Example: double[ ] balance;

6. Classes, objects, methods

Any project created in C# has a class, which is a good example of the encapsulation, inheritance and polymorphism features of object-oriented languages. The syntax of class definitions is not complicated, remember the class keyword, which is the keyword that defines the class.

The specific syntax of class definitions is as follows.

Class access modifiers Modifiers Class name

```
{
```

member of the class

}

(1)The class access modifiers are used to restrict access to the class, and include public, internal or unwritten; internal or unwritten means that the class can only be accessed in the current project; public means that the class can be accessed in any project.

(2)Modifiers: Modifiers are descriptions of the characteristics of the class itself and include abstract, sealed and static. abstract means abstract and a class using this modifier cannot be instantiated; sealed is a sealed class and cannot be inherited; static is a static class and cannot be instantiated.

(3)Class names: Class names are used to describe the function of a class, so it is best to define class names in a meaningful way so that the user can understand what is being described in the class. Class names must be unique under the same namespace.

(4)Members of a class: elements that can be defined in a class, mainly fields, properties and methods.

A class is the most basic C♯ type. A class is a data structure that combines state (fields) and operations (members of methods and other functions) in a single unit. Classes provide definitions for dynamically created instances of classes, which are also called objects. Classes support inheritance and polymorphism, which is the mechanism by which derived classes can be extended and specialised to base classes.

New classes can be created using class declarations. A class declaration begins with a declaration header, which is composed as follows: the attributes and modifiers of the class are specified, followed by the name of the class, last by the base class (if any) and the interfaces implemented by the class. The declaration header is followed by the class body, which consists of a set of member declarations positioned between a pair of curly brackets "{"and"}".

The memory occupied by the object is automatically reclaimed when the object is no longer in use. In C♯, it is not necessary or possible to explicitly free the memory allocated to an object.

7. Brief description

(1) Object: an entity in the real world (everything in the world is an object).

(2) Classes: a collection of objects with similar properties and methods.

(3) Features of object-oriented programming: encapsulation, inheritance, polymorphism.

(4) The three elements of an object: properties (what the object is), methods (what the object can do), and events (how the object responds).

8. Unity Scripting System

Scripting is an essential component of all games.

(1) Unity scripting overview: Although Unity uses the standard Mono runtime implementation for scripting, it still has its own conventions and techniques for accessing the

engine from scripts.

(2) Console Panel, The Console window menu contains options for opening log files, controlling the number of messages displayed in the list and setting up stack tracing. the Console toolbar contains options for controlling the number of messages displayed and for searching and filtering messages(Figure 4-1).

Figure 4-1

①Clear: Removes all messages generated from the code, but retains compiler errors.

②Collapse: Shows only the first occurrence of a duplicate message. This option is useful in the case of runtime errors that are sometimes generated with each frame update (e. g. null references).

③Clear On Play: Automatically clears the console whenever you enter play mode.

④Clear on Build: Clears the console when building a project.

⑤Error Pause: LogError is called from the script and playback is paused.

⑥[Attach-to-Player]: Opens a drop-down menu containing options for connecting to a development version running on a remote device and for displaying its player log in the console.

⑦Player Logging: If the console is connected to a remote development version, this option will enable player logging for the version. Disabling this option will suspend logging, but the console will still be connected to the target version. Disabling this option will also hide the rest of the options in this drop down menu. Select any version listed under Player Logging to display its log in the Console window.

⑧Editor: If the console is connected to a remote development version, select this option to display the logs from the local Unity Player instead of the logs from the remote version.

⑨<Enter IP>: Open the Enter Player IP dialog box where you can specify the IP address of the development version on the remote device.

⑩Connect: Click the Connect button in the dialog box to connect to a version and add it to the list of development versions at the bottom of the drop-down menu.

⑪[DEVELOPMENT BUILDS]: Lists the available development versions. This includes automatically detected versions as well as versions added using the Enter IP option.

⑫Message switch: Displays the number of messages in the console. Click to show/hide the messages.

⑬Warning switch: Displays the number of warnings in the console. Click to show/hide the warnings.

⑭Error switch: Displays the number of errors in the console. Click to show/hide errors.

(3) Creating and using scripts. The behaviour of game objects is controlled by additional components, and Unity allows the use of scripts to create the components themselves. Using scripts you can trigger game events, modify component properties at any time, and respond to user input in any way desired.

Unity negatively supports the C♯ programming language, C♯ (pronounced C-sharp) is an industry standard language similar to Java or C++.

In C♯ Creating scriptsUnlike most other resources, scripts are usually created directly in Unity. New scripts can be created from the Create menu in the top left of the Project panel, or Execute "Assets"→"Create"→"C♯ Script"Command from the main menu(Figure 4-2).

Figure 4-2

Create a script file called First Scripts. This script will open in a text editor when you double click on a script resource in Unity(Figure 4-3).

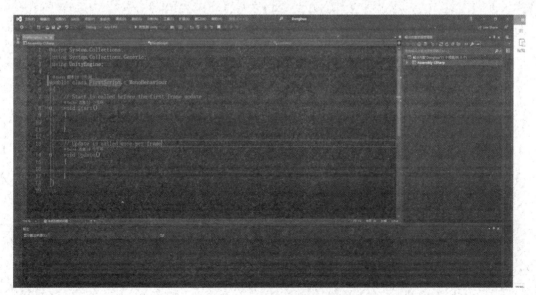

Figure 4-3

The code is as follows:

```
using System. Collections;
using System. Collections. Generic;
using Unity-engine;

public class First Script : Mono-behaviour
{
    // Start is called before the first frame update
    void Start()
    {

    }
    // Update is called once per frame
    void Update()
    {

    }
}
```

Two methods, Start() and Update, are provided by default in the script. These two methods are part of the Unity life-cycle and are both of type private. The Start() method is called on the first frame that the script is active, and the Update() method is really called every frame. The Update function is where the code is placed to handle the frame updates of the game object. This may include moving, triggering actions and responding to user in-put, basically anything that needs to be handled over time as the game runs.

Project 5 The Game Object class in Unity

5. 2 Theoretical guidance

The base class for all entities in the Unity scene. Many variables in the Game Object class have been removed.

There is no object that we use more in Unity development than the Game Object class. Here is a brief write up of some examples of the methods in the Game Object class.

1. Creation of objects

There are many ways to create objects.

(1)Creation via the Game Object menu bar

(2)Creation by code

(3)Drag and drop from resources into the scene

2. Common properties and methods in scripts

(1) activeInHierarchy. Defines whether the Game Object is active in the Scene (Figure 5-1).

GameObject.activeInHierarchy

public bool **activeInHierarchy** ;

描述

定义 GameObject 在 Scene 中是否处于活动状态。

这可使您知道 GameObject 在游戏中是否处于活动状态。如果启用了其 GameObject.activeSelf 属性，以及其所有父项的这一属性，则该 GameObject 处于活动状态。

Figure 5-1

Represents the actual state of the object in the scene(Figure 5-2).

Figure 5-2

(2)active Self. The local active state of this Game Object(read-only). As show in Figure 5-3.

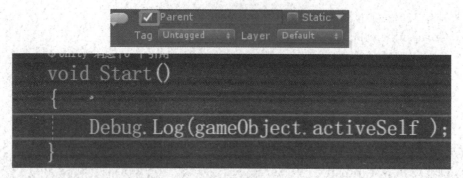

Figure 5-3

Only the local activation status of the object can be display, if you want to know the actual activation status, please use the activeInHierarchy(Figure 5-4).

Figure 5-4

(3)is Static. Editor-only API that specifies whether the game object is static or not (Figure 5-5).

3. Constructors

Can be used to create instances of game objects(Figure 5-6).

(1)Find. Finds a Game Object by name and returns. This function returns only the active Game Object(Figure 5-7).

Figure 5-5

Figure 5-6

GameObject.Find

Figure 5-7

(2)FindGameObjectsWithTag. Returns a list of the active Game Objects tagged with tag. If no Game Object is found，an empty array is returned(Figure 5-8).

Figure 5-8

(3) FindWithTag. Returns an active Game Object tagged with tag, or null if no Game Object is found (Figure 5-9).

Figure 5-9

(4) GetComponent. If the game object has a component of type type attached to it, return it, otherwise return null (Figure 5-10).

Figure 5-10

(5) SetActive. Activates/deactivates the Game Object depending on the given value true or /false/. A Game Object may be inactive because its parent item is not active. In this case, calling SetActive will not activate it, but only set the local state of this Game Object, which can be checked using Game Object. active Self. Unity can use this state when all parent items are active (Figure 5-11).

Figure 5-11

(6) Destroy. Deleting a Game Object, component or resource. The actual object destruction operation is always delayed until the end of the current Update loop, but always completed before rendering (Figure 5-12).

Figure 5-12

(7)Instantiate. Clones the original object and returns the cloned object. This function creates a copy of the object in a similar way to the 'Duplicate' command in the Editor. If the object to be cloned is a Game Object，you can also optionally specify its position and rotation（otherwise，the default is the position and rotation of the original Game Object）. If the clone is a Component，the Game Object to which it is attached will also be cloned，where the optional position and rotation can also be specified. When cloning a Game Object or Component，all child objects and components will also be cloned，with their properties set the same as the original object. The active state of the Game Object will be passed on when cloning，so if the original object is inactive，the clone will also be created in an inactive state（Figure 5-13）.

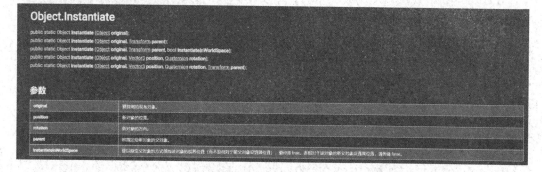

Figure 5-13

Project 6　Transform class in Unity

6.2　Theoretical guidance

1. Overview of the transform component

The transform component determines the value of each object's Position, Rotation and Scale properties in the scene. There is a transform component for each game object.

Figure 6-1

Properties component: Position, Rotation, Scale (Figure 6-1).

2. Transform class (position, rotation and scaling of objects)

Each object in a scene can have a transformation. It is used to store and manipulate the position, rotation and scaling of an object. Each transform can have a parent, allowing you to apply position, rotation and scaling in a hierarchical way(Figure 6-2).

Figure 6-2

(1)Position. The position property of the transform component is a local coordinate relative to the parent object, so if it has no parent, the properties box will naturally show the world coordinates. If it has a parent, then the value displayed in the Unity editor is the coordinate based on the Transform's parent at a higher level. The Transform. position represents the global coordinates of the current Component(Figure 6-3).

Figure 6-3

（2）Rotation and eulerAngles. Rotation in Unity is divided into quaternion（x，y，z，w）and Euler angles（x，y，z）.

Use transform. rotation. z to obtain a Cos，Sin value instead of the panel value. Instead，use transform. eulerAngles Z is a value from 0 to 360. When rotating in the positive direction，you can obtain the corresponding value on the panel，but when rotating in the reverse direction，the value will decrease from 360 all the way down.

Rotation：a quaternion used to store the rotation of the transformation in world space（Quaternion quaternion）（Figure 6-4）.

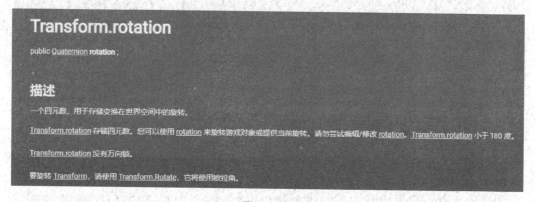

Figure 6-4

（3）EulerAngles. Rotation in Euler angle（in degrees）（Vector3）（Figure 6-5）.

Figure 6-5

3. Native and world

Parenting is a very important concept in Unity. When a game object is the parent of another game object, its child game objects move, rotate and scale with it, just as your arms belong to your body and when you rotate your body, your arms rotate with it. But child objects move, rotate and scale, and the parent object does not change with them. Any object can have more than one child object, but only one parent object(Figure 6-6).

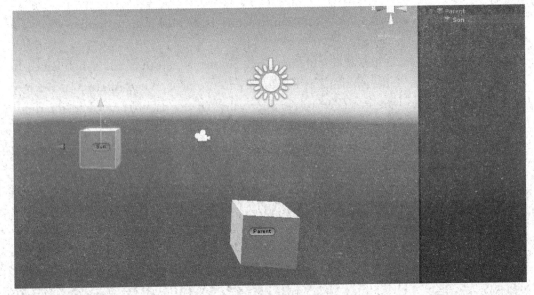

Figure 6-6

localEulerAngles: The rotation in degrees relative to the parent transformation, expressed in Euler angles.

local Position: The position of the transformation relative to the parent transformation.

localRotation: Transform rotation relative to the parent transform rotation.

local Scale: Transform scaling relative to the parent object.

lossy Scale: Global scaling of the object(read-only).

(1)Setting the object's parent object. Expose the properties in the script to expose them to the Inspector panel of the object to which they belong(Figure 6-7). The script code is as follows:

```
using System. Collections;
using System. Collections. Generic;
using Unity-engine;
public class NewBehaviourScript  : Mono-behaviour
{
    public int num=3 ;
    // Start is called before the first frame update
    void Start()
```

```
    {


    }
}
```

（Values are based on the data in the Inspector panel.）

Parent：The parent of the transformation（Figure 6-8）. The script code is as follows：

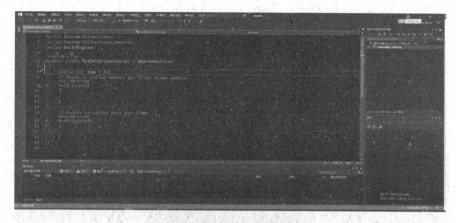

Figure 6-7

```
using System. Collections;
using System. Collections. Generic;
using Unity-engine;
public class NewBehaviourScript  : Mono-behaviour
{
    public Transform parent Tran ;
     // Start is called before the first frame update
     void Start()
     {
          Transform.  parent＝parentTran ;
     }
}
```

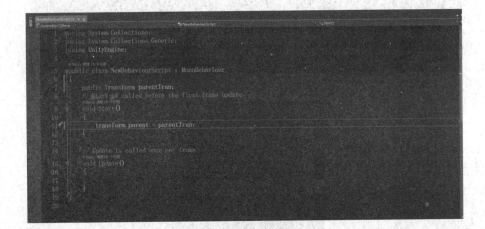

Figure 6-8

(2)Translate(displacement setting)(Figure 6-9).

```
public void Translate(float x, float y, float z);
public void Translate(float x, float y, float z, [DefaultValue("Space.Self")] Space relativeTo);
public void Translate(Vector3 translation);
public void Translate(Vector3 translation, [DefaultValue("Space.Self")] Space relativeTo);
public void Translate(float x, float y, float z, Transform relativeTo);
public void Translate(Vector3 translation, Transform relativeTo);
```

Figure 6-9

Move the transformation according to the direction and distance of the translation(Figure 6-10).

Transform.Translate

public void **Translate** (Vector3 **translation**);
public void **Translate** (Vector3 **translation**, Space **relativeTo**= Space.Self);

描述

根据 translation 的方向和距离移动变换。

如果 relativeTo 被留空或设置为 Space.Self，则会相对于变换的本地轴来应用该移动。（在场景视图中选择对象时显示的 X、Y 和 Z 轴。）如果 relativeTo 为 Space.World，则相对于世界坐标系应用该移动。

Figure 6-10

Project 7 Prefab prefabricated body

7. 2 Theoretical guidance

A Game Object can be created manually through the Create menu under the Hierarchy panel, or new when we want to create a game object dynamically within the program. Game Object, which may contain various settings, components and some scripts. Prefab allows us to use the same object in different Scene, or even Project, for example if I implement a bullet, by packaging it as a Prefab I can use it directly in another game.

1. Creating a prefabricated body

(1)Drag and drop the Cube directly into the Assets panel of the Project panel to create a prefabricated body(Figure 7-1).

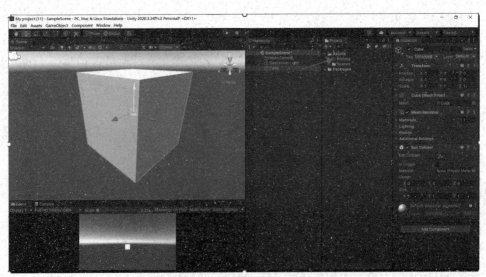

Figure 7-1

（2）You can observe that the Cube turns blue and a Cube appears in the Project panel （Figure 7-2）.

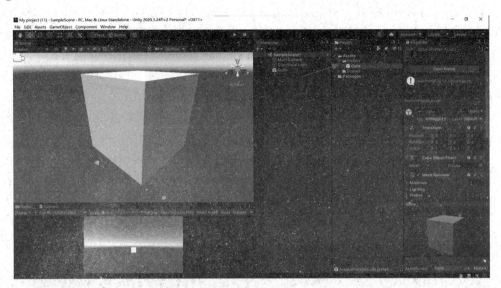

Figure 7-2

（3）Select the Cube in the scene，you can see in the Inspector panel more such a column，click Select to locate the prefabricated body in the project template "master"，click Revert is to restore the current prefabricated body，Apply is to determine the current prefabricated body（Figure 7-3）.

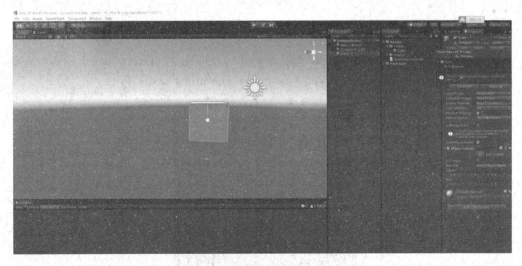

Figure 7-3

2. Instantiate

This function creates a copy of the object in a similar way to the 'Duplicate' command in the Editor. If the object to be cloned is a Game Object，you can also optionally specify its position and rotation (otherwise, the default is the position and rotation of the original Game Object). If the clone is a Component, the Game Object it is attached to will also be cloned, where the optional position and rotation can also be specified. When cloning a Game Object or Component, all child objects and components will also be cloned, with their properties set the same as the original object. By default, the parent of the new object will be null, so it is not 'sibling' to the original object. However, can set the parent object using an overloaded method. If the parent is specified but not the position and rotation, then the position and rotation of the original object is used as the local position and rotation of the clone; if the instantiateInWorldSpace parameter is true, then the world position and rotation of the original object is used. If position and rotation are specified, they are used as the object's position and rotation in world space. The active state of the Game Object is passed on cloning, so if the original object is in an inactive state, the cloned object will also be created in an inactive state(Figure 7-4).

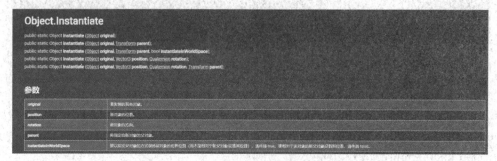

Figure 7-4

3. Instantiation of prefabricated bodies

Drag and drop the prefabricated body directly into the Hierarchy panel. You can see that the prefabricated body has turned blue in the Hierarchy panel(Figure 7-5).

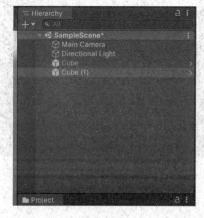

Figure 7-5

Project 8 Collider

8.2 Theoretical guidance

The Collider component defines the shape of the object to be used for physics collisions. Colliders are invisible and their shape does not need to be identical to the object's mesh; in fact, a rough approximation is often more effective and harder to detect during game play.

Unity offers four different Collider components:

(1)Box Collider: Cube Collider, the Cube object mounts this Collider by default;

(2)Capsule Collider: Capsule-shaped Collider, the Capsule object mounts this Collider by default;

(3)Sphere Collider: Spherical Collider, the Sphere object mounts this Collider by default;

(4)Mesh Collider: Collider shaped according to the mesh, the Plane object mounts this Collider by default;

1. Collisions in a nutshell

The simplest (and lowest processor overhead) collision bodies are the so-called primitive collision body types. In 3D, these are box colliders, spherical colliders and capsule colliders. In 2D, 2D box colliders and 2D circular colliders can be used. Any number of these collision bodies can be added to a single object to create a composite collision body.

Collision bodies can be added to objects that do not have rigid components to create floors, walls and other static elements of the scene. These are known as static collision bodies. In general, static collision bodies should not be repositioned by changing the transform position, as this can significantly affect the performance of the physics engine. Collision bodies on objects with rigid bodies are called dynamic collision bodies. Static collision bodies can interact with dynamic collision bodies, but as there are no rigid bodies, they do not respond to collisions by moving(Figure 8-1).

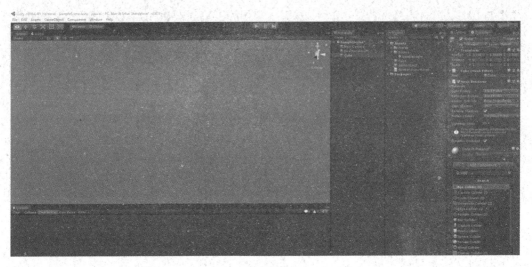

Figure 8-1

2. Introduction to the main scripts

(1) OnCollisionEnter. OnCollisionEnter is called automatically when the collision body/rigid body has started to touch another rigid body/collision body(Figure 8-2).

Figure 8-2

(2)OnCollisionExit. OnCollisionExit is called automatically when the collision body/ rigid body has stopped touching another rigid body/collision body(Figure 8-3).

Figure 8-3

(3)OnCollisionStay. OnCollisionStay is automatically called once per frame for each rigid body/collision that is in contact with it(Figure 8-4).

Figure 8-4

(4)OnTriggerEnter. OnTriggerEnter is called automatically when the Collider other event enters this trigger. This message is sent to the trigger Collider and the Rigidbody to which the trigger Collider belongs (if any), as well as to the Rigidbody or Collider(if there is no Rigidbody) that touches the trigger(Figure 8-5).

Figure 8-5

(5)OnTriggerExit. OnTriggerExit is called automatically when Collider other has stopped touching the trigger(Figure 8-6).

Figure 8-6

(6)OnTriggerStay. OnTriggerStay is called automatically for "almost" all frames of other Colliders that are touching the trigger. this function is located on the physical timer, so it does not have to run every frame(Figure 8-7).

Figure 8-7

Project 9　Rigidbody

9.2　Theoretical guidance

Rigid bodies allow game objects to behave in a physically controlled manner. Rigid bodies can receive forces and torques to move the object in a realistic manner. Any game object must contain a rigid body that is affected by gravity and behaves in a way that is based on applied forces (through scripting) or interacts with other objects through the NVIDIA PhysX physics engine(Figure 9-1).

1. Properties explained

(1) Mass: indicates the quality of the object(default is kg).

(2) Drag: air resistance, the amount of air drag affecting an object when moving it according to force.

(3) Angular Drag: Is the angular resistance, when the object is rotated according to the torque, the air resistance of the object is affected.

(4) Use Gravity: It is called on gravity, if this property is enabled, the object will be affected by gravity (ticked by default).

Figure 9-1

(5) Is Kinematic: It is called dynamic mode on, if this option is enabled, the object will not be driven by the physics engine and can only be manipulated by Transform.

(6) Interpolate: The interpolation only try to use one of the options provided if you see sharp movements in rigid body motion.

①Interpolate: Smoothes the transformation based on the transformation of the previous frame.

②Extrapolate: Smoothes the transformation based on the estimated transformation of the next frame.

(7)None: No interpolation is applied.

(8)Collision Detection: Is called collision detection mode is used to prevent fast moving objects from passing through other objects without detecting collisions.

①Discrete: uses discrete collision detection for all other colliding objects in the scene. Other collision bodies will use discrete collision detection when testing for collisions. Used for normal collisions (this is the default value).

②Continuous: uses discrete collision detection for dynamic collision bodies (with rigid bodies) and sweep-based continuous collision detection for static collision bodies (without rigid bodies). Rigid bodies set to Continuous Dynamic will use continuous collision detection when testing collisions with that rigid body. Other rigid bodies will use discrete collision detection. Used for Continuous Dynamic detection of objects requiring collision (This property has a significant impact on physics performance, so if collision of fast objects is not an issue, leave it at the Discrete setting).

③Continuous Dynamic uses: sweep-based continuous collision detection for game objects set to Continuous and Continuous Dynamic collision. Continuous collision detection will also be used for static collision bodies (no rigid bodies). For all other collision bodies, discrete collision detection is used. For fast moving objects.

④Continuous Speculative: uses speculative continuous collision detection for rigid and colliding bodies. This is also the only CCD mode that can be set for moving objects. This method is usually less expensive than sweep-based continuous collision detection.

(9)Constraints: restrictions on rigid body motion.

①Freeze Position: Is called location freezing, selectively stops the movement of the rigid body along the world's X, Y and Z axes.

②Freeze Rotation: Is called angle freezing, selectively stops rigid body rotation around local X, Y and Z axes.

2. AddForce

Applies force continuously in the direction of the force vector. Force can be specified as ForceMode /mode/ to change the force type to Acceleration, Impulse or Velocity Change. If the GameObject is inactive, AddForce has no effect. By default, once a force is applied (except for the Vector3. zero force), the state of the rigid body is set to wake up.

3. ForceMode properties

ForceMode has the following four attributes(Figure 9-2).

(1)Force: Adds a sustainable force to the Rigidbody, affected by Mass.

(2)Acceleration: Adds a sustainable acceleration to the Rigidbody, ignoring Mass effects.

(3)Impulse: Immediately adds an impulsive force to the Rigidbody, affected by Mass.

(4)VelocityChange: Immediately adds a velocity to the Rigidbody, ignoring Mass effects.

Properties

Force	Add a continuous force to the rigidbody, using its mass.
Acceleration	Add a continuous acceleration to the rigidbody, ignoring its mass.
Impulse	Add an instant force impulse to the rigidbody, using its mass.
VelocityChange	Add an instant velocity change to the rigidbody, ignoring its mass.

Figure 9-2

Project 10　Input System

10.2　Theoretical guidance

Interface to access input systems. Use this class to read the axes set in traditional game inputs and to access multi-touch/accelerometer data on mobile devices. GetAxis with one of the following default axes: "Horizontal" and "Vertical" are mapped to the joysticks (A, D, W, S and arrow keys). "Mouse X" and "Mouse Y" are mapped to mouse increments. "Fire1""Fire2" and "Fire3" are mapped to the Cmd, keys and the three mouse or joystick buttons. New input axes can be added. To use inputs for any type of movement behaviour, use Input. It gives you smooth and configurable inputs, which can be mapped to keyboard, joystick or mouse. GetButton is only used for actions such as events. Input. GetAxis will make the script code much cleaner. It is also important to note that the Input flag is not reset until the Update command is performed. It is recommended that all Input calls are made in the Update loop(Figure 10-1).

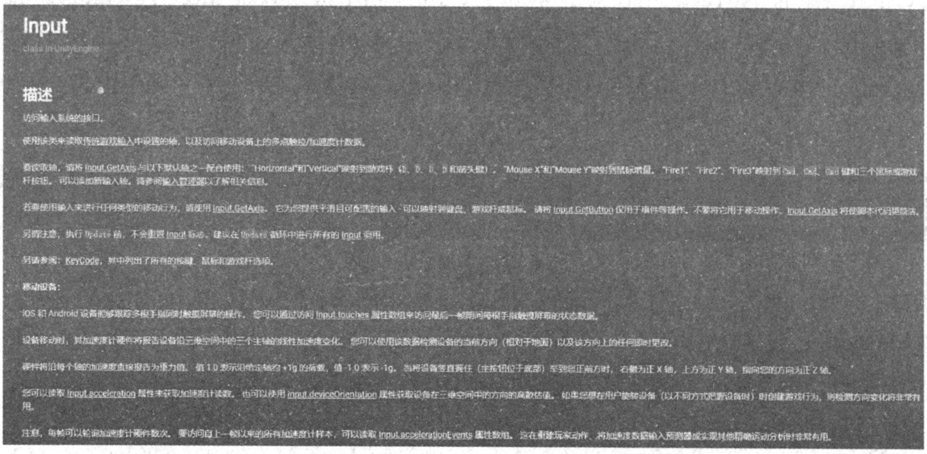
Input

class in UnityEngine

描述

访问输入系统的接口。

使用该类来读取传统游戏输入中设置的轴，以及访问移动设备上的多点触控/加速度计数据。

要读取轴，请将 Input.GetAxis 与以下默认轴之一配合使用："Horizontal"和"Vertical"映射到游戏杆（和箭头键），"Mouse X"和"Mouse Y"映射到鼠标增量，"Fire1"、"Fire2"、"Fire3"映射到键和三个鼠标或游戏杆按钮。可以添加新输入轴。请参阅输入管理器以了解相关信息。

若要使用输入来进行任何类型的移动行为，请使用 Input.GetAxis。它为您提供平滑且可配置的输入 - 可以映射到键盘、游戏杆或鼠标。请将 Input.GetButton 仅用于事件等操作。不要将它用于移动操作。Input.GetAxis 将使脚本代码更简洁。

另请注意，执行 Update 前，不会重置 Input 标志。建议在 Update 循环中进行所有的 Input 调用。

另请参阅：KeyCode，其中列出了所有的按键、鼠标和游戏杆选项。

移动设备：

iOS 和 Android 设备能够跟踪多根手指同时触摸屏幕的操作。您可以通过访问 Input.touches 属性数组来访问最后一帧期间每根手指触摸屏幕的状态数据。

设备移动时，其加速度计硬件将报告设备沿三维空间中的三个主轴的线性加速度变化。您可以使用该数据检测设备的当前方向（相对于地面）以及该方向上的任何即时更改。

硬件将沿每个轴的加速度直接报告为重力值。值 1.0 表示沿给定轴约 +1g 的荷载，值 -1.0 表示 -1g。当将设备竖直握住（主按钮位于底部）举到您正前方时，右侧为正 X 轴，上方为正 Y 轴，指向您的方向为正 Z 轴。

您可以读取 Input.acceleration 属性来获取加速度计读数。也可以使用 Input.deviceOrientation 属性获取设备在三维空间中的方向的离散估值。如果您想在用户旋转设备（以不同方式把握设备时）时创建游戏行为，则检测方向变化将非常有用。

注意，每帧可以轮询加速度计硬件数次。要访问自上一帧以来的所有加速度计样本，可以读取 Input.accelerationEvents 属性数组。这在重建玩家动作、将加速度数据输入预测器或实现其他精确运动分析时非常有用。

Figure 10-1

1. Virtual keys

GetAxis returns the value of the virtual axis identified by axisName, the return value is of type float(Figure 10-2).

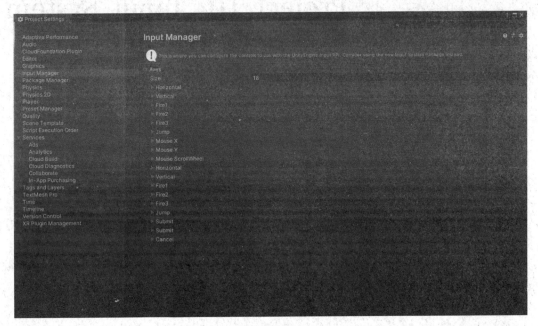

Figure 10-2

For keyboard and joystick inputs, this value will be in the range -1-1. If the axis is set to incremental mouse movement, the mouse increment will be multiplied by the axis sensitivity, in a range other than-1-1.

This value is independent of the frame rate; there is no need to worry about frame rate variation when using this value. To set the input or see the options for axisName, Execute "Edit"→"Project Settings"→"Input"command. this will bring up the Input Manager. expand Axis to see a list of the current inputs. You can use one of these as /axisName/. To rename an input or change a Positive Button etc. , expand one of the options and change the name in the Name field or the Positive Button field. In addition, change the Type to Joystick Axis. to add a new input, add 1 to the number in the Size field(Figure 10-3).

Figure 10-3

2. button script

(1) GetButton: returns true when the virtual button identified by buttonName is pressed and held.

Imagine an automatic firing scenario where the function returns true for as long as the button is held down. Use this function only when implementing an event that triggers an action, such as a weapon firing. The buttonName parameter is usually one of the names in the InputManager, e. g. Jump or Fire1. GetButton returns false when the button is released(Figure 10-4).

Figure 10-4

(2)GetButtonDown: returns true during the frame in which the user presses the virtual button identified by buttonName(Figure 10-5).

Figure 10-5

(3)GetButtonUp: returns true on the first frame of releasing the virtual button identified by buttonName after it has been pressed by the user(Figure 10-6).

Figure 10-6

(4)GetKey: returns true when the user presses the key identified by name(Figure 10-7).

Figure 10-7

(5)GetKeyDown: returns true during the frame in which the user starts pressing the key identified by name(Figure 10-8).

Figure 10-8

(6)GetKeyUp: returns true during the frame in which the user releases the key identified by name after pressing it(Figure 10-9).

Input.GetKeyUp

public static bool GetKeyUp (string name);

Figure 10-9

(7)GetMouseButton: returns whether the given mouse button is pressed.

A button value of 0 indicates a left arrow button on the keyboard, 1 indicates a right arrow button on the keyboard, and 2 indicates an up and down (middle) arrow button on the keyboard. Returns true when the mouse button is pressed and false when it is released (Figure 10-10).

Input.GetMouseButton

public static bool GetMouseButton (int button);

Figure 10-10

(8)GetMouseButtonDown: returns true during the frame in which the user presses the given mouse button.

A button value of 0 indicates the left arrow button on the keyboard, 1 indicates the right arrow button on the keyboard, and 2 indicates the up and down (middle) arrow buttons on the keyboard. Returns true when the mouse button is pressed and false when it is released(Figure 10-11).

Input.GetMouseButtonDown

public static bool GetMouseButtonDown (int button);

Figure 10-11

(9)GetMouseButtonUp: returns true during the frame in which the user releases the given mouse button after it has been pressed.

A button value of 0 indicates the left arrow button on the keyboard, 1 indicates the right arrow button on the keyboard, and 2 indicates the up and down (middle) arrow buttons on the keyboard. Returns true when the mouse button is pressed and false when it is released(Figure 10-12).

Input.GetMouseButtonUp

public static bool GetMouseButtonUp (int button);

Figure 10-12

Project 11　UI system

11.2　Theoretical guidance

The UI system can be used to create user interfaces quickly and intuitively. This section will introduce the main features of the UI system(Figure 11-1).

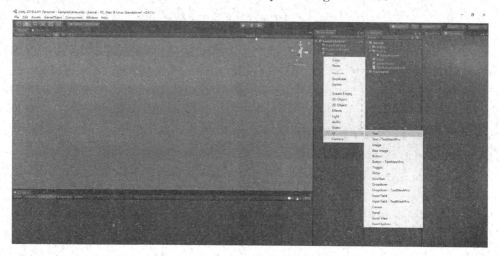

Figure 11-1

1. Summary of Canvas

The Canvas is the area that should hold all UI elements. A canvas is a game object with a canvas component, and all UI elements must be children of the Canvas. When creating a new UI element (e. g. creating an image using the menu GameObject > UI > Image), a canvas is automatically created if there is not already one in the scene, and the UI element is created as a child of this canvas. However, not only one canvas can exist in a scene, so multiple canvases can be used as required.

The canvas area is displayed as a rectangle in the Scene view. This makes it easy to position UI elements without always displaying the Game view. The canvas will use EventSystem objects to assist the messaging system. The role of the EventSystem is to handle the interaction events of the UI in the scene(Figure 11-2).

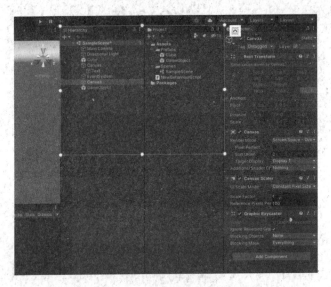

Figure 11-2

2. Render Mode

Canvas has a Render Mode setting that can be used to render in either screen space or world space.

(1)Screen Space - Overlay(the default Render Mode for Canvas is Screen Space - Overlay). This render mode places the UI elements on top of the scene rendered on the screen. If the screen is resized or the resolution is changed, the canvas will automatically be resized to fit this situation. In this mode will not be able to manually resize the Canvas etc. as the size of the Canvas is determined by the size of the Game view, which is overlaid on top of the entire Game view.

(2)Screen Space—Camera. This rendering mode is similar to Screen Space—Overlay, but in this mode the canvas is placed at a given distance in front of the specified camera. UI elements are rendered from this camera, which means that the camera setting affects the appearance of the UI. If the camera is set to an orthogonal view, the UI elements are rendered in perspective view and the amount of perspective distortion can be controlled by the camera field of view. If the screen is resized, the resolution is changed or the camera view cone is altered, the canvas will also be automatically resized to suit this.

(3)World Space. In this rendering mode, the canvas behaves the same as all other objects in the scene. The canvas size can be set manually with the rectangle transform and the UI elements will be rendered in front of or behind other objects in the scene based on their 3D position. This mode is useful for UIs that are meant to be part of the world. This interface is also known as a 'narrative interface'. It is important to note that all Canvas must be set to World Space in the development of VR projects.

　3. UI—text

　　The text component, also known as a Label, has a text area where you can enter the text to be displayed. You can set the font, font style, font size and whether the text supports the subtext feature. There are options to set the alignment of the text, settings for horizontal and vertical overflow (controlling what happens when the text is larger than the width or height of the rectangle) and a Best Fit option to make the text resize to fit the available space.

　　Create text: Right click in the Hierarchy panel, Execute—"UI"→"text"command (Figure 11-3).

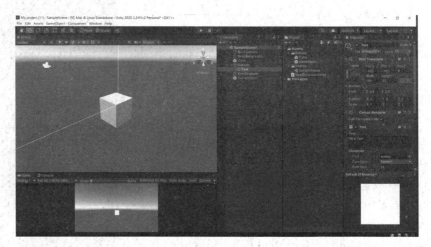

Figure 11-3

There are several main attributes:

(1)TextThe text displayed by the control.

(2)Character.

①Font The font used to display the text.

②Font StyleThe style applied to the text. Options include Normal, Bold, Italic and Bold and Italic.

③Font Size The size of the displayed text.

④Line Spacing The vertical spacing between lines of text.

⑤Rich TextWhether the markup element in the text should be interpreted as a subtext style, which can display the effect of HTML tags. If you are interested in HTML tags, you can look them up online.

(3)Paragraph.

①Alignment: The horizontal and vertical alignment of text.

②Align by Geometry: Performs horizontal alignment using a range of glyph geometries (rather than glyph indicators).

③Horizontal OverflowA method for dealing with cases where text is too wide to fit in-

side a rectangle. The options provided are Wrap and Overflow.

④Vertical Overflow: A method for dealing with cases
where line feed text is too tall to fit within a rectangle. The
options provided are Truncate and Overflow.

⑤Best FitUnity: should ignore the size property and
try to fit the text directly into the control's rectangle.

⑥Color is: used to render the colour of the text.

⑦Material is: used to render the material of the text
(Figure 11-4).

4. UI—Button

The button has an OnClick UnityEvent that defines the
action that will be performed when the button is clicked; the
Button element will have a Text element as a child by default
(Figure 11-5).

(1)Introduction to component properties(Figure 11-6).

① Interactable: Check, the button is available, uncheck, the button is not available.

② Transition: The button's own transition method
when changing state: default is Color Tint, Sprite Swap,
Animation.

③Normal Color: The colour of the initial state.

④Highlighted Color: The selected state or the proxim-
ity of the mouse will bring it to the highlighted state.

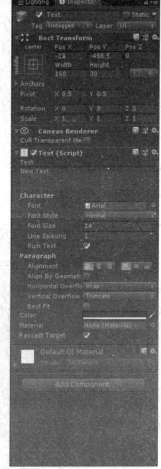

Figure 11-4

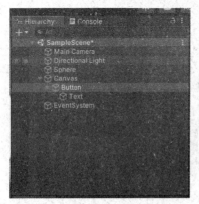

Figure 11-5

Figure 11-6

⑤Pressed Color: Mouse click or press Enter when the button is selected.

⑥Disabled Color: Colour when disabled.

⑦Color Multiplier: The greater the colour switching speed, the faster the colour changes between several states.

⑧Fade Duration：The delay time of the colour change，the greater the change the less noticeable it is.

Sprite Swap，the message that appears can be set like this.

①Highlighted Sprite：Highlighted when selected or when the mouse is near.

②Pressed Sprite：Pressed Enter when the mouse is clicked or the button is selected.

③Disabled Sprite：when the image is disabled.

(2)Keystroke triggering steps are as follows(Figure 11-7)：

Figure 11-7

①Create a public method(Figure 11-8).

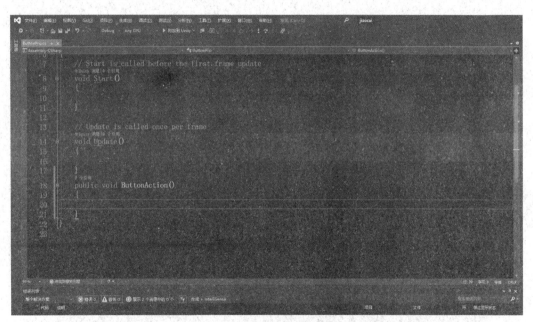

Figure 11-8

②Attaching scripts to game objects in the scene(Figure 11-9).

③Click on ＋ in the options(Figure 11-10).

④Drag the object with the script into the corresponding position(Figure 11-11).

⑤Select the method you have just created in Function(Figure 11-12).

This willtrigger the appropriate method(Figure 11-13).

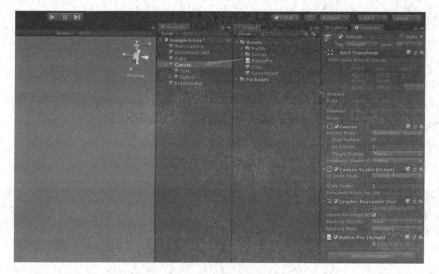

Figure 11-9

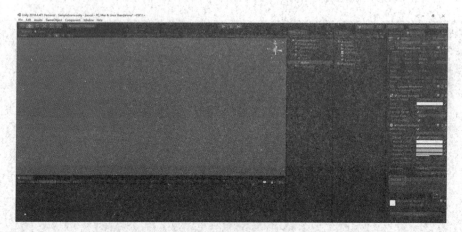

Figure 11-10

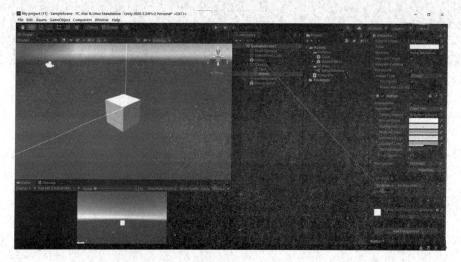

Figure 11-11

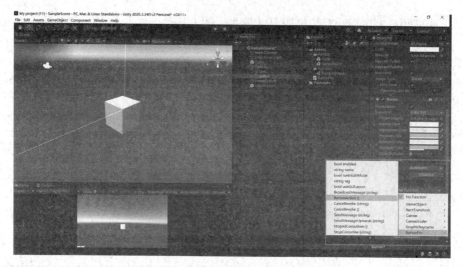

Figure 11-12

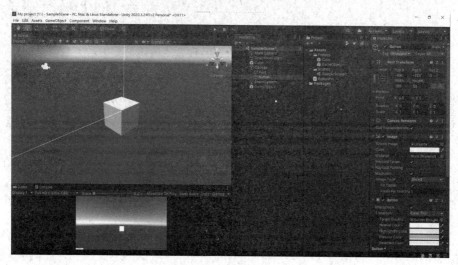

Figure 11-13

5. UI—Image

The Image control displays a non-interactive image to the user. This image can be used for decoration, icons etc. or the image can be changed from a script to reflect changes made to other controls. The control is similar to the Raw Image control, but provides more options for animating the image and accurately filling the control rectangle. However, the Image control requires its texture to be a sprite, whereas the Raw Image can accept any texture.

Introduction to component properties(Figure 11-14).

(1) Source Image：Texture indicating the image to be displayed (must be imported as a sprite).

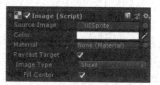

Figure 11-14

(2)Color：The colour to be applied to the image.

(3)Material：The material used to render the image.

(4)Raycast Target：Whether this image should be considered as a ray casting target.

(5)Preserve Aspect：Ensure that the image remains at its current size.

(6)Set Native Size：Use this button to set the size of the image box to the original pixel size of the texture.

6. UI—Panel

A panel control, also known as a panel, is actually a container on which other UI controls can be placed. As you move the panel, the UI controls placed within it move with it, making it easier to move and handle a group of controls in a more logical and convenient way. The panel can be resized by dragging the 4 corners or 4 edges of the panel control (Figure 11-15).

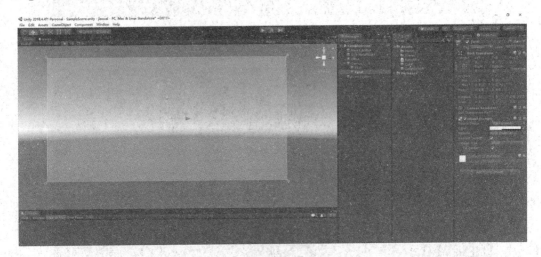

Figure 11-15

7. Toggle

The Toggle element acts as a switch. We can only click on a Button when Toggle is on, otherwise we are not allowed to click on the Button. Add the Toggle element to the Canvas as shown in the Figure 11-16.

The Toggle element has two sub-objects, Background and Label, both of which are used to control the appearance of the Toggle(Figure 11-17).

8. Slider

Slider sliders, which can be used to control volume etc. Add the Slider element to the Canvas as shown in the Figure 11-18.

The Slider element is made up of several image elements and the Slider style can be changed by replacing the image in the Image(Figure 11-19).

9. LayoutGroup

The function of the LayoutGroup component is to control the size and position of the sub-objects. LayoutGroup component consists of Horizontal Layout Group(arrange the child objects horizontally)，Vertical Layout Group(arrange the child objects lengthwise) and Grid Layout Group(grid the child objects)．LayoutGroup components do not control their own layout properties，which can be controlled automatically or set manually using the components mentioned above．LayoutGroup components can be nested in any combination．

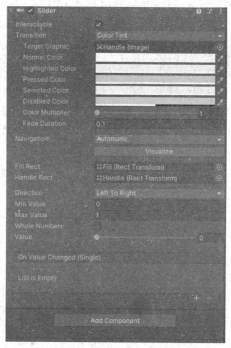

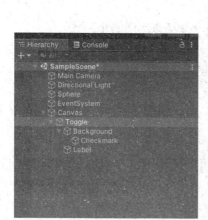

Figure 11-16　　　　　　　　　　　　　　Figure 11-17

Figure 11-18　　　　　　　　　　Figure 11-19

(1)Horizontal Layout Group(arrange the child objects horizontally)(Figure 11-20).

①Padding：Inner Margins.

②Spacing：Spacing between elements.

③Child Alignment：Alignment of child objects.

④Child Controls Size：Whether child objects control their own size.

⑤Child Force Expand：Whether to force the child object to stretch to fill all available space.

(2)Vertical Layout Group(arrange the child objects lengthwise)The rules are similar to those of the Horizontal Layout Group(Figure 11-21).

Figure 11-20   **Figure 11-21**

(3)Grid Layout Group(grid the child objects).

Unlike horizontal/vertical arrange，Grid arrange ignores the minimum，preferred，flexible properties of the sub-objects and uses Cell Size to set their size(Figure 11-22).

①Padding：Inner Margin.

②Cell Size：Size of the sub-object.

③Spacing：Spacing between sub-objects.

④Start Corner：The position of the first element in the layout.

⑤Start Axis：Orientation of the layout，horizontal or vertical.

⑥Child Alignment：The way the child objects are a-ligned to each other.

Figure 11-22

⑦Constraint：Set some constraints on the Grid（number of rows/columns of share-holders）to aid the automatic layout system.

Project 12   Ray

12. 2   Theoretical guidance

A ray is represented as an infinitely long line starting at origin and travelling in a certain direction(Figure 12-1).

(1)Use of rays. The Ray and the RaycastHit collision information class are the two most commonly used ray tool classes.

Figure 12-1

RaycastHit: The structure used to obtain information from a raycast.

It's properties are as follows:

①barycentricCoordinate: The coordinates of the centre of gravity of the hit triangle.

②collider: The Collider of the hit.

③distance: The distance from the origin of the ray to the point of impact.

④lightmapCoord: The coordinates of the UV lightmap at the point of impact.

⑤normal: The normal of the surface hit by the ray.

⑥point: The impact point in world space where the ray hit the collider.

⑦rigidbody: The rigidbody of the hit collision body, or null if the collision body is not attached to a rigid body.

⑧textureCoord: The UV texture coordinate of the collision location.

⑨textureCoord2: The secondary UV texture coordinate at the point of impact.

⑩transform: The transform of the hit rigid body or collision body.

⑪triangleIndex: The index of the triangle hit.

(2)Emission method(Physics. Raycast). Throws a ray to all colliding bodies in the scene, starting at origin, facing the direction of the direction, with a length of maxDistance (Figure 12-2).

Figure 12-2

(3) The detailed code is as follows(Figure 12-3):

```
using System.Collections;
using System.Collections.Generic;
using UnityEngine;

public class RayPro   : MonoBehaviour
{
Ray ray;
RaycastHit hit;
void Start()
{
ray= new Ray(this.transform.position. Vector3. forward);
  }
    // Update is called once per frame
  void Update()
  {
      If(Physics.Raycast(ray.out hit))
      {
        Debug.Log(hit.transform.name);
}
    }
}
```

Figure 12-3

Project 13 Pathfinding System

13.2 Theoretical guidance

(1)The steps to create a navigation grid are as follows:

①Select the scene geometry that should affect navigation: walkable surfaces and obstacles.

②Check the Navigation Static checkbox to include the selected objects in the navigation mesh baking process(Figure 13-1).

③Adjust the baking settings to match the agent size. Among them: Agent Radius defines how close the centre of the agent is to a wall or window sill; Agent Height Defines how low the agent can reach; Max SlopeDefines how steeply the agent can walk up a ramp; Step Height Defines how high an obstacle the agent can step on(Figure 13-2).

Figure 13-1

Figure 13-2

④Click Bake to build the navigation grid(Figure 13-3).

(2)NavMesh Agent. The Nav MeshAgent component helps you create characters that can avoid each other as they move towards their target. The Agent uses the NavMesh to infer the game world and knows how to avoid each other and other obstacles to movement. Pathfinding and spatial inference are handled using the NavMesh Agent's scripting API (Figure 13-4).

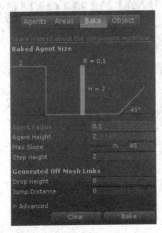

Figure 13-3   Figure 13-4

①Agent Size: Agent Type.

②Base offset: The offset of the collision cylinder with respect to the centre of the transformation axis.

③Steering include: Speed the maximum speed of movement (expressed in world u-nits/second); Angular Speed maximum speed of rotation in degrees/sec; Acceleration maximum acceleration (expressed in world units/square second); Stopping distance the a-gent will stop when the distance to the target position reaches this value; Auto Braking when this property is enabled, the agent will slow down when reaching the target. This property should be disabled for behaviour such as patrolling (in which case the agent should move smoothly between multiple points).

④Obstacle Avoidance include: RadiusThe radius of the agent, used to calculate colli-sions between the obstacle and other agents. HeightThe height clearance required for the agent to pass through the overhead obstacle. Quality of obstacle avoidance, if you have a large number of agents, you can save CPU time by reducing the quality of obstacle avoid-ance. If avoidance is set to none, only collisions will be resolved and no active avoidance of other agents and obstacles will be attempted. PriorityWhen performing obstacle avoid-ance, this agent will ignore agents with lower priority. The value should be in the range 0-99, where a lower number indicates a higher priority.

⑤Path Finding include: Auto Traverse Off-Mesh Link Set to true to automatically traverse off-mesh links. This should be turned off if off-mesh links are to be traversed u-sing animation or in a particular way. Auto RepathWhen this property is enabled, the a-gent will try to find its way again when it reaches the end of a partial path. When no path to the target is reached, a partial path is generated to the nearest reachable location to the target. Area Mask describes the type of area that the agent will consider when pathfinding. Each grid area type can be set when preparing the grid for navigation grid baking. For example, stairs can be marked as a special area type and stairs can be disabled for certain character types.

(3)Basic method of pathfinding(NavMeshAgent)(Figure 13-5).

①SetDestinationsets or updates the target to trigger a new path calculation. Set or update the destination to trigger a new path calculation. Note that paths may not be available until several frames later. Path-

Figure 13-5

Pending will be true when the path is calculated. If a valid path is available, the agent will resume moving(Figure 13-6).

Figure 13-6

②destinationGets the agent's destination in the world coordinate system units or attempts to set the agent's destination within them. If the destination is set but the path has not yet been processed, the location returned will be the closest valid navigation grid location to the previously set location; If the agent has no path or requested a path, the position of the agent on the navigation grid is returned; If the agent is not mapped to a navigation grid (for example, there is no navigation grid in the scene), the position at infinity is returned(Figure 13-7).

Figure 13-7

(4)Creating Off-Mesh Links. Off-Mesh Links are used to create paths through the outside of a walkable navigable mesh surface. For example, jumping over a ditch or fence, or opening a door before passing through it, can all be described as off-mesh links. We will add an Off-Mesh Link component to describe a jump from the upper platform to the ground(Figure 13-8).

(5)Nav Mesh Obstacle. The component allows you to describe moving obstacles (for example, barrels or crates controlled by a physical system) that the Nav Mesh Agent should avoid when navigating through the world. When an obstacle is moving, the Nav Mesh Agent will try to avoid it. When the obstacle is stationary, it will carve a hole in the navigation grid. The navigation grid agents will then change their path to go around the obstacle or find a different route if the obstacle causes the path to be completely blocked

(Figure 13-9).

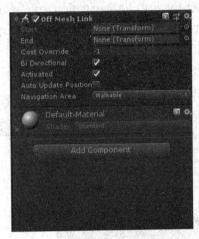
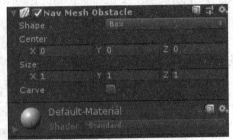

Figure 13-8   Figure 13-9

①Shape: The shape of the obstacle geometry, select the option that best fits the shape of the object. There are two shapes called Box and capsule. So Box needs to be set up: Center the centre of the box (relative to the transformation position) and the size of the box. Capsule needs to be set up Center the centre of the capsule (relative to the transformation position), Radius the radius of the capsule and Height the height of the capsule.

②Carve: When the Carve checkbox is checked, the navigation grid obstacle creates a hole in the navigation grid.

③Move Threshold: Unity treats a navigation grid obstacle as moving when it moves a distance beyond the value set by Move Threshold. Use this property to set this threshold distance to update the moving carved hole.

④Time To Stationary: The amount of time (in seconds) to wait for an obstacle to be treated as stationary.

⑤Carve Only Stationary: When this property is enabled, the obstacle will only be carved when it is at rest.

Project 14 Animations

14.2 Theoretical guidance

Unity's animation features include redirectable animations, full control of animation weights at runtime, event calls during animation playback, complex state machine hierarchical views and transitions, blended shapes for facial animation and much more.

1. Overview of the animation system

Unity has a rich and complex animation system (sometimes called "Mecanim"). The system has the following capabilities:

(1) Simple workflow and animation settings for all elements of Unity including objects, characters and properties.

(2) Support for imported animation clips and animations created within Unity.

(3) Human animation redirection - ability to apply animations from one character model to another.

(4) Simplified workflow for aligning animation clips.

(5) Easy preview of animation clips and the transitions and interactions between them. As a result, animators and engineers work more independently of each other, allowing animators to prototype and preview animations before hooking into the game code.

(6) Provides visual programming tools to manage the complex interactions between animations.

(7) Animation of different body parts with different logic.

(8) Layering and masking capabilities.

2. Animation workflow

Unity's animation system is based on the concept of animation clips, animation clips contain information about how certain objects should change their position, rotation or other properties over time. Each clip can be considered as a single linear recording. Animation clips from external sources are created by artists or animators using third-party tools such as Autodesk © 3ds Max © or Autodesk © Maya ©, or from motion capture studios or other sources. The animation clips are then programmed into a structured flowchart-like system called the Animator Controller, which acts as a 'state machine', keeping track of which clip should currently be played and when the animations should be changed or blended together.

A very simple Animator Controller might contain just one or two clips, for example, using this clip to control energy blocks rotating and bouncing, or to set the animation for opening and closing doors at the correct time. A more advanced Animator Controller may contain dozens of humanoid animations for all of the protagonist's movements, and may mix between multiple clips at the same time to provide smooth movement as the player moves through the scene.

Unity's animation system also has a number of special features for working with humanoid characters. These features allow humanoid animation to be redirected to your character model from any source (e. g. motion capture, Asset Store or some other third-party animation library), and muscle definition can be adjusted. These special features are enabled by Unity's stand-in system; in this system, humanoid characters are mapped to a common internal format. All these components (Animation Clip, Animator Controller and Avatar) are attached together to a game object via the Animator component. The component references the Animator Controller and (where necessary) the Avatar of the model, which in turn contains a reference to the animation clip used.

3. Animation Controller

(1) Animator component. The Animator component is used to assign animations to the game objects in the scene; the Animator component needs to refer to the Animator Controller, which defines which animation clips are to be used and controls when and how blending and transitions are made between animation clips. If the game object is a humanoid character with an Avatar definition, an Avatar should also be assigned in this component as follows Figure 14-1.

①Controller: The Animator Controller attached to this character.

②Avatar: The Avatar of this character (if the Animator is used to animate a humanoid character)

③Apply Root Motion: Whether we should control the position and rotation of the character from the animation itself or from the script.

Figure 14-1

④Update Mode: This option allows you to choose when the Animator should be updated and which time scale should be used.

⑤Normal Animator: Updates in sync with the Update call and the speed of the Animator matches the current time scale. If the time scale slows down, the animation will match by slowing down.

⑥Animate Physics Animator: Updated in sync with the FixedUpdate call (i. e. , at the same pace as the physics system). This mode should be used if you want to animate the movement of objects with physical interactions (e. g. characters that can push rigid objects around).

⑦Unscaled Time Animator: Updates in sync with the Update call, but the speed of the Animator ignores the current time scale and animates at 100% speed regardless. This option can be used to animate the GUI system at normal speed while using the modified time scale for effects or pausing the game.

⑧Culling Mode: You can select the culling mode for the animation.

⑨Always Animate: Always animate and do not cull even when off-screen.

⑩Cull Update Transforms: Disables redirection, IK (inverse dynamics) and writing of transform components when the renderer is not displayed.

⑪Cull Completely: Disables animations completely when the renderer is not displayed.

(2)Creating an Animator Controller. Character behaviour can be viewed and set from the Animator Controller view (execute "Window"→"Animation"→"Animator" command). Animator Controllers can be created in a number of ways.

①From the Project view, execute"Create"→"Animator Controller"command.

②Right-click in the Project view and execute "Create"→"Animator Controller"command.

③From the Assets menu, execute "Assets"→"Create"→"Animator Controller"comand. (Figure 14-2).

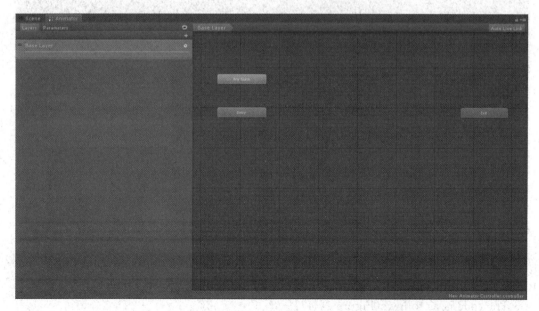

Figure 14-2

4. Animated Transitions

State machine transitions can help you simplify large or complex state machines. It allows for a higher level of abstraction of state machine logic. Each view in the Animator window has an Entry and Exit node. These nodes are used during state machine transitions. The Entry node is used when transitioning to a state machine. The Entry node will be evaluated and will branch to the target state according to the conditions set. In this way, the entry node can control the initial state of the state machine by evaluating the state of the parameters when the state machine is started.

Because the state machine always has a default state, there will always be a default transition from the entry node branching to the default state.

Animation parameters are variables defined in the Animator Controller that can be accessed from the script and assigned values to. This is how the script controls or influences the flow of the state machine. For example, the value of a parameter can be updated via an animation curve and then accessed from a script so that the pitch of the sound effect can be changed (as if it were an animation). Similarly, scripts can set the values of parameters that are picked up by Mecanim. For example, a script can set parameters to control the blend tree.

Default parameter values can be set using the Parameters section of the Animator window (selectable in the top right corner of the Animator window). These parameters can be divided into four basic types. Float(a number with a decimal part); Int(integers); Bool (true or false value, represented by a checkbox); Trigger(Boolean parameter that is reset by the controller when used by a transition, represented by a round button)(Figure 14-3).

Figure 14-3

5. Program control

The following functions in the Animator class can be used to assign values to parameters from scripts: SetFloat, SetInt, SetBool, SetTrigger and ResetTrigger.

The code is as follows(Figure 14-4):

```
void Update()
    {
        float h= Input.GetAxis("Horizontal");
        float v= Input.GetAxis("Vertical");
        bool fire= Input.GetButtonDown("Fire1");
        animator.SetFloat("Strafe",h);
        ananimator.SetFloat("Forward",v);
        aimator.SetBool("Fire",fire);
}
void OnCollisionEnter(Collision col)
{
        if( col.gameObject.CompareTag("Enemy"))
    {
            animator.SetTrigger("Die");
    }
}
```

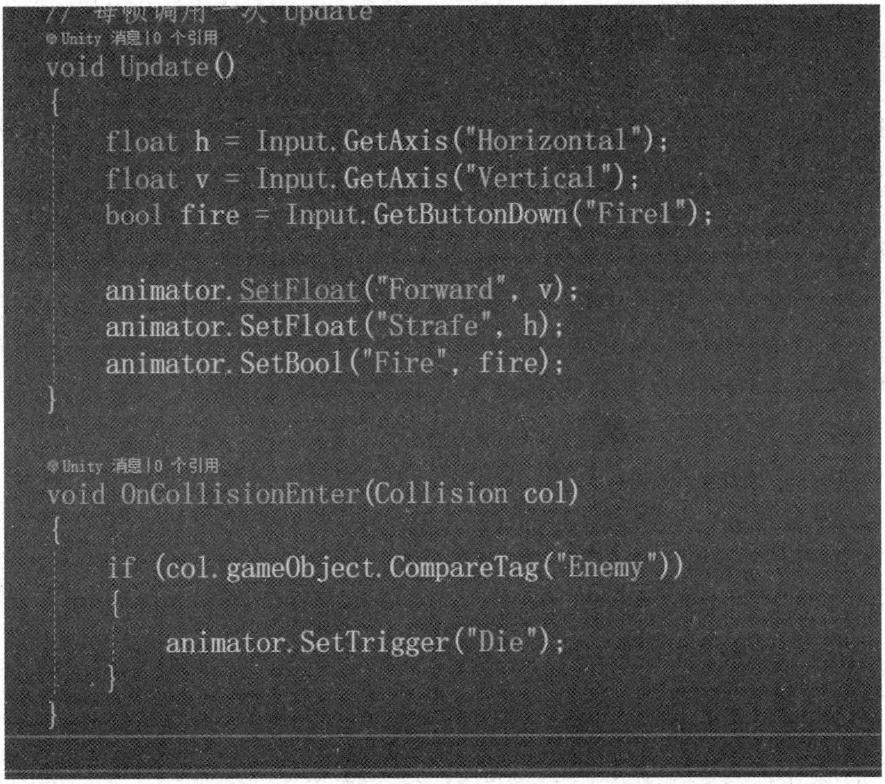

Figure 14-4

6. Animation layers

Unity uses animation layers to manage complex state machines for different body parts. For example, you have a lower body layer for walking/jumping, and an upper body layer for throwing objects/shooting. You can manage the animation layers from the Layers widget in the top left corner of the Animator Controller(Figure 14-5). Click on the gear on the right hand side of the window to display the settings for that layer(Figure 14-6).

Figure 14-5

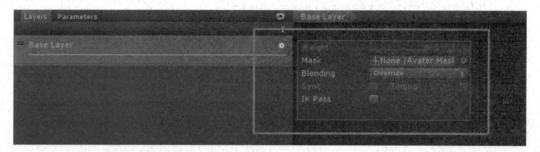

Figure 14-6

On each layer you can specify the mask (part of the animation model to which the animation is applied) and the blend type. override means that information from other layers will be ignored, while Additive means that the animation will be added on top of the previous layer (a new layer can be added by pressing + above the widget). The Mask property is used to specify the mask to be used on this layer. For example, if you want to play a throw animation for only the upper half of the model, while allowing the character to also walk, run or stand still, you could use a mask on the layer so that the throw animation is played at the location where the upper part of the body is defined.

7. Animation

The Animation window in Unity allows you to create and modify animation clips directly from within Unity. It is designed to act as a powerful and straightforward alternative to external 3D animation programs. In addition to animating motion, the editor also allows you to animate variables for materials and components and to enrich animation clips with animation event functions (which are called at specified points on the timeline).

(1) Introduction to the Animation view. The Animation view is used to preview and edit animation clips of game objects that have been animated in Unity. To open the Animation view in Unity, execute"Window"→"Animation"command(Figure 14-7).

Figure 14-7

(2) View animations on game objects. The Animation window is associated with the Hierarchy window, the Project window, the Scene view and the Inspector window. Like the Inspector, the Animation window displays the animation timeline and keyframes for the currently selected game object or animation clip resource. You can use the Hierarchy window or Scene view to select a game object, or the Project window to select an animation clip resource.

(1)Properties.

①Animation timeline(Figure 14-8). On the right side of the Animation view is the timeline of the current clip. The keyframes for each animation property are displayed in this

Figure 14-8

timeline. The timeline view has two modes: Dopesheet mode and Curves mode. To switch between these modes, click on Dopesheet or Curve at the bottom of the animation properties list area.

②Timeline(Figure 14-9). The keyframe list mode provides a more compact view, allowing

Figure 14-9

you to view the keyframe sequence for each attribute in a single horizontal track. As a result, you can view an overview of the keyframe times of multiple attributes or game objects.

③Controls for playback and frame navigation(Figure 14-10).

Figure 14-10

From left to right, these controls are: Preview mode (toggle on/off); Record mode (toggle on/off) Note, if record mode is on, preview mode will also always be on; Move the playback head to the beginning of the clip; Move the playback head to the previous keyframe; Playback animation; Move the playback head to the next keyframe; Move the playback head to the end of the clip.

Playback can also be controlled using the following keyboard shortcuts: Press a comma","to jump to the previous frame; Press the full stop"." to jump to the next frame; Hold Alt and press the comma","to jump to the previous keyframe; Hold Alt and press the full stop"." to jump to the next keyframe.

8. BlendTree

A common task in game animation is to blend between two or more similar movements in order to make the movement between characters more natural and fluid. Perhaps the most familiar example is the blending of walking and running animations depending on the speed of the character. Another example is when a character leans to the left or right when turning during a run.

It is important to distinguish between Transitions and Blend Trees. Although both are used to create smooth animations, they are used in different kinds of situations.

Transitions are used to smoothly transition from one animation state to another in a given amount of time. Transitions are specified as part of the animation state machine. Transitions from one motion to a completely different motion are usually not a problem if the transition is quick.

The blend tree allows for smooth blending of animations by combining multiple animations to varying degrees. The effect of each motion on the final effect is controlled by a blend parameter, which is just one of the digital animation parameters associated with the Animator Controller. In order for the blended motion to make sense, the motions to be blended must have similar properties and timing. A blend tree is a special type of state in an animation state machine.

(1)Right-click on a blank area on the Animator Controller window.

(2)From the context menu that is displayed, execute "Create State"→"From New Blend Tree"command.

(3)Double click on the Blend Tree to go to the Blend Tree Graph.

The Animator window now shows a graphical representation of the entire Blend Tree, while the Inspector shows the currently selected node and its direct children(Figure 14-11).

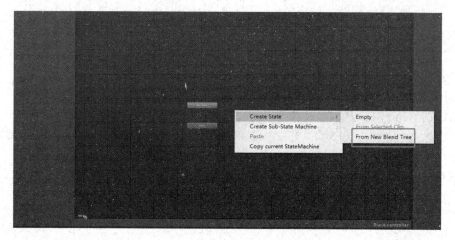

Figure 14-11

To add an animation clip to the blend tree, select it and then click the plus icon in the Motion field of the Inspector(Figure 14-12).

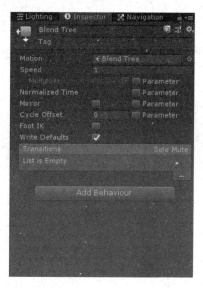

Figure 14-12

参 考 文 献

[1] Unity 公司. Unity 2017 从入门到精通[M]. 北京：人民邮电出版社，2020.

[2] 薛庆文. Unity 3D 从入门到精通[M]. 北京：电子工业出版社，2021.

[3] 吴亚峰，索依娜，于复兴. Unity 案例开发大全[M]. 2 版. 北京：人民邮电出版社，2018.

[4] [澳]杨栋. 创造高清 3D 虚拟世界：Unity 引擎 HDRP 高清渲染管线实战[M]. 北京：电子工业出版社，2021.